THE SIX SIGMA
PERFORMANCE
HANDBOOK

Other Books in the Six Sigma Operational Methods Series

THE SIX SIGMA PERFORMANCE HANDBOOK

A Statistical Guide to Optimizing Results

Praveen Gupta

McGraw-Hill

New York   Chicago   San Francisco   Lisbon   London   Madrid
Mexico City   Milan   New Delhi   San Juan   Seoul
Singapore   Sydney   Toronto

The McGraw·Hill Companies

Library of Congress Cataloging-in-Publication Data

Gupta, Praveen
    The Six Sigma performance handbook / Praveen Gupta.
        p.   cm.
    ISBN 0-07-143764-9
    1. Six sigma (Quality control standard)   2. Quality control—Statistical methods.   3.
    Production management—Statistical methods.   I. Title.

    TS156.G869   2004
    658.4'013—dc22

                                                                        2004055945

1 2 3 4 5 6 7 8 9 0   DOC/DOC   0 1 0 9 8 7 6 5 4

ISBN 0-07-143764-9

*The sponsoring editor for this book was Kenneth McCombs, the
editing supervisor was Caroline Levine, and the production supervisor
was Sherri Souffrance. It was set in Fairfield per the NBF design by
Deirdre Sheean of McGraw-Hill Professional's Hightstown, N.J.,
composition unit.*

Printed and bound by RR Donnelley.

This book is dedicated to the people who have contributed to my continual learning:

My team of Gupta siblings
Veena, Chandra, Sita, Gita, Madhu, Meena, Ishwar, Nami Saran, Kam, Uma, and Pradeep

My colleagues, friends, and mentors
Frank Brletich, Curt Wyman, Roshan Goel, Brian Peterson, Gary Westerman, Jit Lodd, Ashok Patel, Jim Harrington, Jay Patel, and Gard Ayers

The DMAIC pioneers
Mikel Harry, Ph.D., and Richard Schroeder for elevating Six Sigma to a greater height.

CONTENTS

FOREWORD

by T. M. KUBIAK

Over the last two to three years, a plethora of books has been written, each eschewing the virtues of Six Sigma and promising to be the only book you'll ever need. Many authors have been so bold as to state that their book is suitable for a wide variety of audiences including executives, Champions, Black Belts, Master Black Belts, Green Belts, and whatever other color of belts you may have. Because these books attempt a "one-size-fits-all-audiences" approach, they often fall short of their objectives.

Before I even began to read Praveen's book, my cynicism surfaced and I expected to read just another book on Six Sigma. Surely, I would see the same old rehash of well-known material, with no new insight to be gained. Was I surprised!

While the technical aspects of Six Sigma are interesting and challenging, it is the nontechnical issues that create the biggest problems. And, more often than not, the Six Sigma leader or professional is not equipped to address them. More than likely, leaders and professionals alike have risen through the quality ranks and have transitioned naturally over to Six Sigma. *The Six Sigma Performance Handbook: A Statistical Guide to Optimizing Results* is a practical book on Six Sigma authored by an expert and his contributors. There are new concepts, ideas, and other inclusions not seen elsewhere.

Praveen makes a nice cut at separating the tools used by Green Belts, Black Belts, and executives. The book helps the reader to not lose site of the fact that we don't need to train Green Belts or Black Belts. Recognizing this undoubtedly will ease your burden in terms of time, energy, and expense. Besides, Praveen and his contributors have focused on the basic tools that, if used with diligence and rigor, will provide substantial process improvement gains and bottom-line results. The underlying message is basics, basics, and basics! Your leadership will appreciate getting the biggest bang for their buck.

Optimizing Six Sigma is a concept that has not been discussed elsewhere. Think about it. Optimizing an improvement approach, tool, methodology, or whatever you choose to call it. In this case, we see the unique approach of applying the Theory of Constraints to Six Sigma. Remember, like anything else within an organization or company, Six Sigma is comprised of processes that must not stagnate. They, too, must be continually improved in a rigorous and well-defined manner.

I know there's an on-going "chicken-or-the-egg" debate focused on answering the question, "Should you Lean out your processes first and then reduce variation, or vice versa?" To speed up Six Sigma, the Six Sigma professional needs to be well versed in the tools, techniques, and methodologies of each. *The Six Sigma Performance Handbook* nicely integrates key "Lean" tools in the Six Sigma Methodology.

The Six Sigma Performance Handbook eases the reader's burden of implementing Six Sigma as well as articulating it to others (i.e., particularly those nasty naysayers) through an impressive compilation of clear and discernable figures, graphics, and diagrams. Forms are provided to help kick-start the reader's implementation process. With the above said, I am sure you will enjoy reading and learning from this book as much as I did.

T. M. Kubiak
Senior Vice President
Quality and Productivity
Customer Relationship Management
Bank of America

FOREWORD

by LONNIE ROGERS

Praveen Gupta and I have worked together for over two years. I came to know about his Six Sigma experience through Darren Stinson, one of my managers. Given Praveen's conviction about Six Sigma and his help in implementing Six Sigma at my company, Ideal Aerosmith, I am certain that if anyone can write a good book about the subject, Praveen can. His knowledge comes from his association with Six Sigma from its inception and his desire to make it a better tool for continual improvement in the long term.

The Six Sigma Performance Handbook is all about how to implement Six Sigma cost effectively by using simple tools effectively for powerful results. My leadership team has taken Praveen's Six Sigma class and has come away awakened with all the tools he has given them. These same tools have been compiled in *The Six Sigma Performance Handbook*. Praveen's style of writing is like a teacher patiently presenting a subject in the classroom. He makes sure his audience understands a difficult subject like statistics without a lot of pain.

I am excited that Praveen has added another book to the list of those he has authored. As in *Six Sigma Balanced Scorecard*, in which he offered a model for corporate Sigma level, he has developed a number of new tools in this book that will make my life and my staff's life easier at Ideal. I am sure Six Sigma practitioners will find tools such as the Six Sigma Audit Checklist, Six Sigma Project Forms, and their guidelines indispensable.

My staff and I have benefited from Praveen's insights. I believe that readers will benefit equally from insights captured in *The Six Sigma Performance Handbook* on their journey to achieving Six Sigma performance.

Lonnie Rogers
President
Ideal Aerosmith, Inc.
East Grand Forks, Minnesota

PREFACE

None of us had realized that Six Sigma would be the talk of the town 15 years after its creation. I am fortunate to have been part of the Six Sigma journey. It's amazing to see what people can do with a tool for creating value. It's just a matter of understanding the process. Similarly, I believe that the development of statistical thinking is as important as the knowledge of statistical tools. *The Six Sigma Performance Handbook* is an attempt to develop statistical thinking, using statistical tools, to achieve better results and breakthrough improvement.

Experts believe that Six Sigma, like any other invention, will hit a wall unless it undergoes continual improvement. Therefore, Six Sigma must evolve and become better, faster, and cheaper—tasting its own medicine of improvement. Implementation of the Methodology has hit a wall in the sense that companies are mired in developing Black Belts, commiting to Six Sigma without proper planning, and thus failing to realize breakthrough benefits. These are signs of a maturing methodology.

The Six Sigma Performance Handbook has been designed to break the impasse one may encounter with Six Sigma and assist practitioners in using the Methodology in an optimum way to maximize benefits. The continual improvement and enhancement process will apply to the book itself in coming years, based on readers' feedback. Comments and recommendations are welcomed at praveen@qtcom.com.

Praveen Gupta

ACKNOWLEDGMENTS

Writing a book requires the collaboration of many enthusiastic individuals. *The Six Sigma Performance Handbook* is no exception. When I started writing this book, it apeared to be an impossible task—similar to a marathon in that it would require a great deal of preparation and conditioning. But just like a marathon, one needs to first get started, which is half the battle. Then, replenishments are needed in order to continue with perseverance and commitment. Spectators on the roadside provide the emotional uplift needed to carry on when one feels tired.

Assembling this handbook required a great deal of preparation and energy just to start the writing marathon. But it is a tremendous help when the marathon becomes a relay race. *The Six Sigma Performance Handbook* had many contributors who made the handbook possible, as follows:

CHAPTER	CONTRIBUTOR(S)
2. Balanced Approach to Planning	Patricia (Patty) Barten
4. Identifying the Problems	Arvin Srivastava and Marjorie Hook
5. Understanding the Scope of the Problem	Rajiv Varshney
6. Developing a Solution	Rajesh Tyagi
7. Breakthrough Solutions	Arvin Srivastava
8. Sustaining Breakthrough	Shan Shanmugham
9. Optimizing Six Sigma	Shan Shanmugham
10. Speeding up Six Sigma	Kam Gupta and Arvin Srivastava
11. Frequently Asked Questions	Mahender Singh
Appendix: Six Sigma Books	Priya Ponmudi

All the contributors have worked hard to get their chapters ready on time. Without their help, the marathon might not have been completed. I would also like to acknowledge Patty, Arvin, and Shan for their superb effort and unconditional support throughout the development of the book.

By now my family members have become used to me burning the midnight oil regularly. I would like to acknowledge their constant support.

The book would not have been possible without the dedication of Kenneth McCombs, McGraw-Hill Senior Editor, to the idea of optimizing Six Sigma. Ken has inspired me to write this book to help Six Sigma practitioners become more successful. I thank Ken's team members, who have been very supportive throughout the production process and have played a critical role in making this book a reality.

Thanks to Lisa Hamburg and Mike Buetow for publishing my earlier work, on which this book builds, in *PC FAB* and *Circuits Assembly* magazines, published by UP Media.

Finally, I feel fortunate to be part of the Six Sigma journey and to be able to share my experience with readers of my books.

INTRODUCTION

Six Sigma has become an industry for driving improved growth and profitability worldwide. Consulting companies, quality departments, trade associations, universities, websites, chat rooms, and conference organizers are spreading the word like wildfire. However, new research must be conducted to improve the methodology, and the principle of better, faster, and cheaper must be applied to Six Sigma, itself.

I have been involved with Six Sigma since 1986 when I was a wafer quality engineer at Motorola's Semiconductor Product Sector. During those days, Six Sigma was in its development stage. Work was being performed to look beyond Three Sigma; Six Sigma was just an idea, and no one dreamed of it becoming a corporate DNA, culture, or standard of excellence. It was simply going to be Motorola's methodology for improvement. Well, it has turned out to be much more than a methodology—for many businesses today, it is everything.

Learning from my experience with Six Sigma at various companies, I see three categories of implementation.

1. *All out,* where the CEO is excited about the initiative and is passionately pushing Six Sigma in the entire theater of operation.
2. Where Six Sigma has become a methodology for improvement, but the CEO is not involved. Six Sigma has been relegated to the quality department, as was the case with Total Quality Management (TQM).
3. The company is trying half-heartedly to do something with Six Sigma—trying to seek some type of Six Sigma certification (get me the paper!).

In addition to these three modes of implementation, there are companies that are still wondering about Six Sigma but who are hesitating to take action. Others think it's for "big" companies or are intimidated by the statistics and think it's beyond their capabilities.

We all know that the only constant is "change," and Six Sigma is all about a lot of improvement, very fast. Many companies that are implementing Six Sigma have not devised a system to measure its effectiveness. One can debate and ask, "As long as the company is profitable and making money, why bother about knowing the sigma level?" In every age of industrialization, whenever companies have been profitable, they have done so by questioning the need to do it right. We have learned to measure what we value. Therefore, in order to sustain the Six Sigma initiative, one must measure its Sigma level. Otherwise, the Six Sigma initative is purposeless and without a target. Given that no official system exists to measure a corporatewide Sigma level, I have developed the one that is incorporated in this book.

Another issue companies face in the implementation of Six Sigma is cost. They are not used to shelling out several thousand dollars to get someone trained who may not even stay with the company after completing the training. Besides, they find it hard to believe that someone can master the knowledge it takes to become a Black Belt in just four to five weeks. I have heard stories about Black Belts certified either through the training or by taking the examination, who are not happily producing value as expected: about $500,000 to $1 million per year. The Six Sigma Metholodgy must become more affordable and it must be improved. It must be optimized like any other process in a company. This book is based on the premise that the Pareto Principle must be applied to Six Sigma Methodology. This is a hard pill to swallow, but it must be done.

However, all Six Sigma Tools are equally applied in the field. Some tools are "vital" and some are "trivial." Experience has shown that many simple graphical, project management, and leadership tools are used more frequently than many of the advanced statistical tools. For many processes, Mapping, Pareto, and Fishbone work wonders. (Bob Galvin, former Motorola Chairman and CEO, told his people to map the process and fix it.) In order to optimize the methodology, we must maximize the benefits of the Six Sigma tools. Figure I.1 delineates the layout of *The Six Sigma Performance Handbook.*

Chapters 1 to 3 look into current issues companies are facing in implementation of Six Sigma and establish a baseline of Six Sigma practices. The planning for Six Sigma must examine various aspects of management, including leadership commitment, employee involvement, and identifying areas that can benefit from application of the Six Sigma Methodology: Strategic Planning, Change Management, and Sustaining Six Sigma. Chapter 3 covers the roadmap for implementing Six Sigma successfully, including tools that are incorporated in the body of

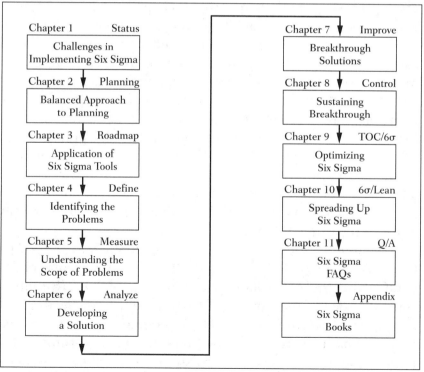

FIGURE I.1.   The Six Sigma Performance Handbook layout.

knowledge for Black Belts, Green Belts, and even executives. One of the key aspects of this chapter is the inclusion of Six Sigma project forms for each phase of the DMAIC (Define, Measure, Analyze, Improve, and Control) Methodology. This chapter looks into the innovative aspect of Six Sigma, which is its defining tenet for realizing great improvement very quickly.

Chapters 4 to 8 cover the basic set of tools of the DMAIC methodology that are commonly used in solving a variety of problems. Following the 80:20 rule, the book looks into the intent of each phase of the DMAIC methodology and the corresponding tools.

The author's belief is that the development of statistical thinking is much more critical than learning all the tools incorporated in the Methodology. Six Sigma's contribution has been to develop a capacity for Six Sigma thinking in making decisions. Many practitioners overemphasize data-driven decisions that lead to excessive focus on collecting data for various activities during the project. Facts do have their place, however: knowledge of the process, the ability to understand the process based on the facts, and making decisions based on variation are critical aspects of the Six Sigma Methodology. An attempt has been made to explain these tools in as practical a way as possible.

The Six Sigma Performance Handbook emphasizes the Control phase of DMAIC, which is vital to sustain the Six Sigma initative at the corporate as well as the project level. Many new concepts, such as Scorecard, knowledge mangement, integration of Six Sigma in audits, corrective actions, and management review processes, have been incorporated. A Six Sigma audit checklist has been developed to audit the Six Sigma efforts in a corporation.

Chapter 9, "Optimizing Six Sigma," looks into improving the Methodology. Eliyahu Goldratt's Theory of Constraints (TOC) and the Methodology's own tool, SIPOC (Supplier, Inputs, Process, Outputs, and Customers), have been examined to improve the Six Sigma Methodology. The Six Sigma process has been critically investigated for opportunities for improvement. As a result, a list of success factors has been presented in the form of a short to-do list.

Chapter 10, "Speeding Up Six Sigma," offers an efficient integration of Lean tools into the Six Sigma Methodology. A comparative analysis of Lean and Six Sigma is presented and key Lean tools have been discussed for implementation. The object is how to achieve great improvement quickly using Value-Stream Mapping, with a focus on Muda's and 5-S. Key Lean tools and Six Sigma thinking along with the DMAIC methodology can do wonders for an organization.

Finally, Chap. 11, "Frequently Asked Questions," utilizes the concept of Web technology and compiles a set of questions for the curious reader. Resources for the reader have been compiled in the Appendix.

THE SIX SIGMA
PERFORMANCE
HANDBOOK

CHALLENGES IN IMPLEMENTING SIX SIGMA

SIX SIGMA EVOLUTION

This chapter establishes a common understanding of the Six Sigma journey to date and looks into the challenges companies and Six Sigma itself face today. Since its inception in 1986, Six Sigma has really grown by leaps and bounds, from a few processes within Motorola divisions to implementation by many divisions at several thousand companies today. In the leadership and vision given by Bob Galvin, then CEO of Motorola, to Larry Bossidy, Jack Welch, and many more CEOs, Six Sigma has been credited as a most successful business performance improvement system. Starting out as a process improvement methodology, Six Sigma has become a superior strategy that is welcomed as a savior at many companies.

From scratch sheets and loose notes to about 200 books on Six Sigma is tremendous progress. Millions of copies of Six Sigma books have been sold worldwide, many companies have implemented Six Sigma, tens of thousands of certified Black Belts have been produced, and many symposiums and conferences have been held. Six Sigma has come a long way. In less than 20 years, Six Sigma has overrun any known business methodology in world recognition, acceptance, and implementation. And the good news is getting better. Following a typical S-curve, Six Sigma has covered less than a third of its life cycle; it has come out of the developmental stage and is moving into the common use and reproduction stage, before being eventually phased out or taken to next stage of evolution. Analysis of currently published business books shows that about 50 percent offer basic introduction to Six Sigma under a variety of names; 20 percent deal with Six Sigma tools; 15 percent address strategy, scorecards, team dynamics, and leadership; 10 percent cover design for Six Sigma; and only the remaining 5 percent are difficult to categorize as anything to do with Six Sigma.

It has taken 17 years, 10 years since the inventor Bill Smith left his legend behind, for Smith to be recognized as the father of Six Sigma in widespread publications. It has taken almost the same length of time for Motorola to rediscover Six Sigma, its own recipe for profits. However, while Motorola was ignoring Six Sigma, the rest of the world was exploiting its power. Companies such as Allied Signal, Honeywell, Raytheon, Texas Instruments, General Electric, Citibank, Bank of America, Caterpillar, John Deere, Seagate Technology, Du Pont, Ford, and Dow Chemical have benefitted from implementing Six Sigma. The most prominent of them all is General Electric, where Jack Welch, former CEO, gave Six Sigma a second birth with lots of energy. And although Bill Smith and Motorola invented Six Sigma, Mikel Harry contributed toward its direct link to profitability—and that is the driver. So, Motorola made the car, and Mikel drove it to its limits. Of course, the pioneers who implemented Six Sigma at Motorola were Bob Galvin (the leader), Bill Wiggenhorn (the preacher), Bill Smith (the creator), and people like myself (the followers), in mid- to late 1980s.

Now, I look around and say Wow! What Six Sigma has become, from what it was! Bill Smith must have imagined it exactly like this. He must have had some power of vision beyond that of a normal person, to see a universal problem and develop such a powerful methodology.

I had been one of those privileged to observe and participate in the development of the Six Sigma methodology. During some sunny winter break, Bill and I would discuss how to establish a sigma level for a publication process, or we would talk in his office about how to count opportunities for old systems that were so large that no one knew how many parts they had, or we would have conference room debates on how to calculate sigma levels for a group of processes. I remember our four Small Wins to Six Sigma, the first four projects in the communications sector group that I happened to track. It was an interesting exercise, because even if we improved a lot (as we thought in those days), the sigma level would not change appreciably. Sometimes, we would wonder what to do to improve the sigma level, whether to improve the process or increase the opportunities. Of course, the path of least resistance would be to count a few more opportunities; however, we chose the path of challenging ourselves to improve the processes.

MOTOROLA'S EXPERIENCE

Bob Galvin's inspiration and incentives, and Motorola's core values, were the key to achieving great results. As we learned, the long-term success achieved with positive behaviors is much more valuable than the short-term success gained by any means. Well, every step to the achievement of Six Sigma was not a rosy one. We had our tough times, especially in the monthly operations review meetings. I remember middle managers trying their best to prepare for those meetings and still returning with a feeling of falling short. All felt beaten up by demands for more improvement.

Motorola's first two initiatives were Six Sigma Quality and Total Cycle Time reduction. In other words, Six Sigma and Lean thinking were an integral part of Motorola's initiative to improve performance. Six Sigma and Lean are like two sides of the coin, inseparable, and must be practiced together. Six Sigma without Lean practices will not be able to achieve the necessary improvement, and Lean alone without Six Sigma will also not be able to achieve the necessary quality levels. Six Sigma improves effectiveness, while Lean improves efficiency of a process.

Before launching Six Sigma, I remember, various initiatives were announced to us annually. Nobody ever knew how well they worked. Every year, management would get excited about such initiatives, which employees could just glance over. We saw some improvement, but nothing appreciable. We really started feeling Six Sigma personally when we all received the training and were asked to set stretch goals for improvement. Setting the stretch goals was an excellent exercise, as it forced us to think about improvement in our area. During this process, our managers made sure that the goals were employee-driven rather than directed by management.

THE CHAIN REACTION

Looking through my notes, I had made some attempt to go beyond three sigma in the semiconductor manufacturing area, where the number of process steps exceeded 100 and a yield level of 99.73 for each process would result in a maximum yield of 69 percent. With that kind of limitation, we had to look for yields better than 99.73 at each process. The Six Sigma development started in the 1985–1986 time frame. This means Bill must have been thinking about it long before we all came to know about it at work.

The Six Sigma S-curve timeline (Fig. 1.1) shows how Six Sigma has mushroomed over time. As Six Sigma itself has become a growing industry consisting of consultants, publishers, trainers, authors, conferences, trade shows, and software providers, there has come a challenge. Now we have derivatives of Six Sigma approaches; a variety of training programs, consultants, and software programs; different understandings of Six Sigma; various implementations of Six Sigma. We may almost need to apply Six Sigma to the Six Sigma industry to reduce variability in various aspects of Six Sigma. The American Society for Quality (ASQ) certification attempts to standardize the basic understanding of Six Sigma; however, the scope of its contents is huge. It is impossible for a person who has been certified as a Black Belt, but has no significant industrial experience, to practice Six Sigma. On the other hand, it is possible for people who have taken the 5-week training program to comprehend all the concepts and become proficient in

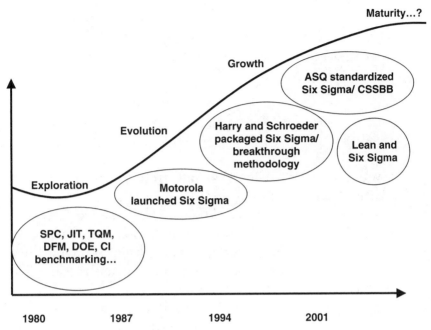

FIGURE 1.1. Six Sigma S-curve.

applying Six Sigma principles. In statistical terms, if a large population is trained in Six Sigma, there will be a small percentage that will become good at applying Six Sigma, while a large percentage of the population will become knowledgeable about Six Sigma tools but be unable to practice the principles effectively, and a smaller percentage of the trained population will have to figure out what to do with the Black Belt training.

POTENTIAL FOR FAILURE

In a recent message board on a website, someone asked for help on Six Sigma in the service area. The timing could not have been better for the writing of this book! One of the many responses included the following:

Number of employees	More than 10,000
Number of Yellow Belts	Most employees

Number of Green Belts	About 2,000
Number of Black Belts	About 500
Number of Master Black Belts	Over 100
Total cost of implementation	15 million
Status	Six Sigma discontinued
Number of Six Sigma experts released	All Black Belts, Master Black Belts, and Champions

There was plenty of advice available from peers and consultants without understanding the root cause.

At another company, I heard (because no one would openly admit it yet) that about 1,200 Black Belts were trained by the leading (i.e., price leader) Six Sigma consulting company. Their CEO, the friend of a Six Sigma legend, created a Six Sigma department and had Six Sigma messages posted throughout the company, on very good looking bulletin boards. During my visit, the company only had the Six Sigma department labels left, and some Black Belts struggling to add value. In the end, I was told that although some projects had been completed successfully, the Six Sigma initiative was about to be folded.

Afterward I attempted to track or gage actual results in various companies. In one of the ASQ local section meetings, someone sought help in identifying projects for Six Sigma after spending more than $1 million on training. In another case, the reported savings of $700,000 was based on some past experiment or process improvement effort. At yet another company where Six Sigma did not produce the desired outcome, management decided to apply Design for Six Sigma and delay the final failure. In companies where the Six Sigma initiative is floundering, the problem is the lack of commitment from the leadership. It is the lack of will to make it work.

After hearing such super success stories about the Six Sigma, one would become curious to learn more about the methodology as well as the results. Surveys have been and are being conducted regularly to better understand perceptions and actual results with a larger sample size. The surveys show that the use of Six Sigma methodology is increasing. There is a debate over whether Six Sigma is a methodology, a strategy, or a culture.

Surveys show that about 25 percent of respondents have been implementing Six Sigma. The number of companies implementing Six Sigma has grown by an order of magnitude. The number of companies implementing Six Sigma is split between manufacturing and nonmanufacturing companies. The companies implementing Six Sigma are quite uniformly distributed by company size.

The benefits of Six Sigma have been limited to a few projects at most companies; otherwise, the Six Sigma initiative does not even go beyond a lot of training. The benefits range from project-specific savings to companywide savings. However, there are no reported data relating the cost of Six Sigma to an organization's profitability and growth. The new Six Sigma Business Scorecard was developed to drive profitability and growth through the Six Sigma initiative.

Implementing Six Sigma in a few large corporations creates a general perception that Six Sigma is not suitable to small businesses. It has been reported that small businesses do not have the resources to implement Six Sigma, implying that Six Sigma is a cost, instead of a system to improve profitability. One of the distinguishing factors for implementing Six Sigma is to improve profitability in short term and facilitate growth in the long term. The initial fear of small- to medium-sized companies is the cost of training. They perceive that getting people trained will cost a lot of money, and then what? This demonstrates that only the methodology aspect of Six Sigma has been understood. For an effective implementation and maximum benefit, Six Sigma must be implemented in its entirety—strategy, methodology, and intent.

Many failures such as the following have been reported.

LEADERSHIP FAILURES

- Not taking enough time to understand and see the benefits of Six Sigma
- Lack of passionate commitment to achieving dramatic results
- Setting low or no expectations for achievement with Six Sigma

- Not treating Six Sigma as a leadership initiative and giving it the highest priority
- Lack of organizational alignment to involve all functions and to change the way of doing things
- Lack of identification of opportunities for improvement relating to profitability before launching Six Sigma
- Not linking compensation structure to actual savings resulting from Six Sigma mindset
- Lack of employee recognition and participation
- Not driving out fear (driving out bottom 10 percent can not be sustained and produces negative behaviors)
- Not changing the product development methodology; design and manufacturing blaming each other
- Not making time for Six Sigma projects, somehow perceived as extra work that must not be important
- Lack of involvement by president or CEO, total delegation of Six Sigma to the corporate champion

BLACK BELT FAILURES

- Limited understanding of leadership and interpersonal aspects of Black Belt role
- Too much emphasis on statistics and complicated tools
- Lack of facilitation skills, causing poor teamwork
- Not clearly defining problem in enough detail
- Lack of innovation through collaboration and systems thinking
- Too many Black Belts, disproportionate to identified opportunities
- Fabricated savings from projects through creative math

GREEN BELT FAILURES

- Not empowered to produce great results
- Low expectations of Green Belts
- Lack of involvement of Green Belts

- Rivalry among Green and Black Belts due to poorly defined roles
- Assignment of Green Belts to supporting roles instead of leading roles
- Perceived lack of respect for Green Belts compared to Black Belts
- Lack of recognition of Green Belts

EMPLOYEE FAILURES

- Failure to enlist employees enough in problem-solving activities
- Lack of an established process for encouraging employees' intellectual involvement
- Insufficient training given to employees in problem solving
- Lack of time given to employees for their active involvement
- Productivity emphasized over creativity and quality
- Lack of empowerment to identify, prioritize, and improve processes

CONSULTANT FAILURES

- Understanding Six Sigma as a collection of advanced and complicated statistical tools
- Focusing on 3.4 parts per million and thinking one has to produce millions of parts to measure sigma level
- Treating Six Sigma like another problem-solving methodology, giving it flavors of "another TQM"
- Not emphasizing the strategic component of Six Sigma that requires the CEO's passionate commitment
- Focusing on training lots of people before identifying the opportunity for profits

CORPORATE FAILURES

Companies that have successfully implemented the Six Sigma initiative have enjoyed significant savings and national and international recognition. However, the well-known company-wide success stories are small in number. Corporate failures have not become known yet, although they have happened. In one instance, a company that had supposedly trained over 1,000 black belts and created a Six Sigma department achieved a limited success, but was unable to sustain it. Later, the Six Sigma department was abandoned, and uncertainty loomed over the status of Six Sigma in that company. Remnants of Six Sigma are still hanging around without any intent of practice.

It has often been remarked that since Motorola invented Six Sigma, why has the company suffered so significantly since about 2000? Observing Motorola's performance in the marketplace both as a former employee and as an interested outsider, I believe that Motorola's challenges have not been primarily due to Six Sigma, although Six Sigma might have accelerated the company's growth, which became difficult to manage. Instead, changing market conditions, changes in leadership, strategic mistakes, and ignorance of Six Sigma contributed to Motorola's woes. Under the leadership of CEO Bob Galvin, Motorola flourished with Six Sigma. But the company was unable to sustain the Six Sigma momentum that fueled its growth engine in the 1987–1992 era.

Polaroid is another company that was touted as a Six Sigma success story and later experienced business challenges. According to Harry and Schroeder in their book *Six Sigma: The Breakthrough Management Strategy Revolutionizing the World's Top Corporations*, (2000), Polaroid committed to implement Six Sigma by 2001, using the Breakthrough Strategy. After many initial successes, Polaroid stumbled in the competitive environment. Today, one can not even find a Six Sigma word on its website. All traces of Six Sigma have been removed. Why? Did Six Sigma cause its failure? During its implementation of Six Sigma, just like Motorola, Polaroid could do no wrong. Did the Six Sigma breakthrough process overburden resources? Or did Six Sigma divert attention from the business sense of making money? Some extremities must have crept in that disturbed the balanced approach required to sustain the Six Sigma journey.

SIX SIGMA EXPERIENCE

Raytheon, in contrast to Motorola and Polaroid, embedded Six Sigma as a vehicle to improve productivity, achieve business growth, reduce cost, and renew its culture. Raytheon has recognized many benefits, including people-to-people networks beyond the Six Sigma projects, customer value, and knowledge-based process management. As noted in its literature, Raytheon has trained over 20,000 employees and completed over 6,000 projects.

Du Pont used Six Sigma to improve productivity; Ford uses Six Sigma to solve its quality problems; Bombardier uses Six Sigma to work better, faster, and more efficiently; GE applied Six Sigma concepts to change its way of doing business; and Motorola developed Six Sigma to renew the corporation and accelerate the improvement process.

Interestingly, early implementers of Six Sigma, such as Motorola, ABB, Texas Instruments, and Polaroid, have struggled. After initial success, these corporations could not sustain either the effort or the performance. In most such cases, something dramatically changes that disrupts the implementation of Six Sigma. The CEO who made Six Sigma successful departs for a new venture, or adventure; a new vice president for quality or Six Sigma leader arrives; explosive growth occurs that underprioritizes the Six Sigma effort; or some catastrophic problem occurs that requires excessive attention. For example, when a company successfully implements Six Sigma, it achieves growth as well as profitability. With the growth come the growing pains. That's when some companies forget about the Six Sigma commitment and waver from it, along with the new employees.

When companies only implement Six Sigma on some projects, it is a very temporary commitment. In such circumstances, Six Sigma has been used mainly as a problem-solving method through application of the Define, Measure, Analyze, Improve, and Control (DMAIC) methodology. However, due to a lack of commitment from the top management, commitment to DMAIC is not sustainable and eventually fades away with the failure of one or more projects.

Many companies that inquire about implementing Six Sigma gather the courage to call some consulting firm to find about the cost. They are practically afraid of implementing Six Sigma because they have heard that companies like GE, Motorola, and other large corporations have saved billions of dollars, and spent millions of dollars. To them, Six Sigma is for a different league, even though its principles sound applicable to their business. Experientially, the consulting firm typically starts with serious Six Sigma training, leading to a significant cash layout for the company, without any sound planning to recoup the expenses. Worse, the money invested in training an employee to become a Black Belt is lost when the Black Belt leaves the company due to lack of commitment to Six Sigma.

The departed Black Belt might as well be the top Quality professional in the company. The new Quality leader is hired. After the initial Six Sigma shock, the company management wants to do something different, this time anything other than Six Sigma. Or, even more disruptive to the Six Sigma process, after some successful years at the company, the CEO leaves for a better opportunity. Departure of the chief executive is a real shocker to the Six Sigma program and all employees involved in Six Sigma. The current team becomes demoralized and the new leadership team arrives and makes its presence felt, even if that requires undoing some of the past good work.

CURRENT IMPLEMENTATION

Some of the new implementations of Six Sigma have been at the Caterpillar, 3M, and other corporations.

CATERPILLAR

Caterpillar launched its Six Sigma initiative simultaneously at all its plants worldwide in 2001. The company has shown remarkable resilience in the current downturn and credits its growth to Six Sigma, with a significant impact on its bottom line. One of Caterpillar's executives has mentioned in a conference that in one quarter, about 60 percent of the company's profitability came from Six Sigma methods. The audience was told that the first-year gains exceeded first-year deployment costs. Caterpillar has about 2,000 Black Belts and 15,000 Green Belts who have been working on over 15,000 projects at a success rate of about 95 percent.

Caterpillar started with DMAIC projects and moved into Define, Measure, Explore, Develop, and Implement (DMEDI), a DFSS equivalent methodology. The projects included process improvement, product improvement, or even improvement of the tools used to do the work. Excited with the success of its Six Sigma initiative, Caterpillar is now extending the initiative to its supply chain. Suppliers and dealers are asked to implement Six Sigma to accelerate continuous improvement. Caterpillar is offering training and implementation assistance to its dealers and suppliers, ensuring their success as well.

3M

At 3M, Six Sigma doctrine was launched in December 2000 and was quickly accepted by the more than 70,000 employees. The company has committed to Six Sigma as a way of doing business. CEO W. James McNerney Jr. says, "We're betting our performance on Six Sigma. This is saying that, if Six Sigma doesn't succeed, the company doesn't succeed." The cost savings have been directly linked to Six Sigma and other performance initiatives. Quick increases in productivity and elimination of waste have been attributed to Six Sigma.

3M is taking the idea of Six Sigma to its customers and suppliers, much in the way Motorola did in the late 1980s. The company is identifying projects with its customers, such as Ford, Wal-Mart, and Home Depot, for new venues of cost saving. 3M expects to save hundreds of millions of dollars using Six Sigma.

SEARS

Interestingly, a search for information about Six Sigma at Sears leads to very limited resources. It appears that Six Sigma is not so publicized at Sears. Six Sigma does not seem to be mentioned in the company's vision or mission statements, and the savings realized at Sears have not been attributed to the Six Sigma initiative, even though the company has deployed quite a few Black Belts to its Six Sigma initiative and requires Green Belt or Black Belt in recruiting new employees.

One might consider that the Sears Six Sigma initiative is not so visible because the company's customers are people who really do care for superior customer service and good value (price and quality). That may have led to emphasis on implementing Six Sigma internally and through its suppliers. Sears is working on several projects internally, without much fanfare.

BANK OF AMERICA

Bank of America has become the next Six Sigma Cinderella. Milton Jones, leader of the Six Sigma initiative at Bank of America stated in a recent article (*Six Sigma Forum,* February 2004) that the bank has saved more than $2 billion in about 3 years, and improved customer satisfaction by 25 percent since 2001. Great performance!

Just as Six Sigma became the DNA of GE, the culture of Motorola, and the standard of excellence of Allied Signal, the initiative has become a combination of business performance metric, approach, and philosophy for Bank of America. As a service business, the bank uses Lean and Six Sigma tools to achieve process improvements in hundreds of projects to increase revenue and customer satisfaction and reduce credit losses and frauds. Now, Six Sigma experience and certification are required in recruitment and promotions.

IDEAL AEROSMITH—SIX SIGMA AT A SMALL COMPANY

One of the challenging questions people ask is how Six Sigma works at small companies. Among all the hyped-up stories of success at large companies, the successes of small companies are little known. One such company is Ideal Aerosmith, located in East Grand Forks, Minnesota. The company provides motion control test products and services to the aerospace and autotest industries.

Ideal started its Six Sigma journey in October 2002 after its president, Lonnie Rogers, listened to Jack Welch's story on audiotape during a long drive. After an executive overview, Rogers communicated his commitment to his management team and empowered them to implement Six Sigma for sustainable growth and profitability. He just believed that Six Sigma was the way to go to realize benefits similar to those larger corporations had gained. Ideal serves customers such as Honeywell and Rockwell.

Ideal established its new corporate vision and goals, conducted Six Sigma awareness training for all employees, performed a business opportunity analysis to identify key projects, and implemented the Six Sigma Business Scorecard. This scorecard includes measurements for the president, managers, and employees in various departments. For each measurement, aggressive improvement goals were set to achieve Six Sigma–level performance. The vision and corporate measurements facilitated the culture for improvement and created a common purpose. Once the measurements were set up, Green Belt training was conducted for managers, who then worked on top-priority projects for improving the process performance, achieving higher customer satisfaction, and realizing corporate growth and profitability objectives.

The first two projects identified by the business opportunity analysis were in the materials and engineering areas. At the corporate sigma level, key process measurements were monitored weekly and monthly. After about 9 months, the company owners felt that they had seen the best performance by the management team. At the next quarterly meeting, the owners felt that the team had exceeded their expectations.

Rogers feels that this is only the beginning of their Six Sigma journey. The team must accelerate the improvement, achieve its full potential, and the reap benefits of the Six Sigma investment. Rogers feels that Six Sigma played a major role in improving the company's profitability, and will help in achieving the planned improvements in quality, customer satisfaction, business growth, and profitability.

SIX SIGMA AND LEAN

When Six Sigma was launched at Motorola, Cycle Time Reduction (so-called Lean) was also launched simultaneously. However, Six Sigma Quality came first. The logic was to do it right first, then do it faster. The reason given was that if we did it faster before we learned to do it right, we would create waste faster. However, people have since combined Six Sigma and Lean into Lean Six Sigma–like solutions. Their focus is on implementing Six Sigma along with the Lean tools.

Figure 1.2 shows the Lean–Six Sigma matrix. The matrix demonstrates that conventional processes lack aspects of both Lean and Six Sigma improvements. The current scenario is based on doing a job. Six Sigma requires doing a job well, and Lean requires doing a job fast. The combination of Six Sigma and Lean drives excellence and efficiency together, which affects quality, productivity, and profitability.

Lean is a way to specify value, arrange value-creating actions in the optimum sequence, conduct these activities without interruption, and improve continually. Lean thinking is a way to do more with fewer resources in providing customers exactly what they want. Value is defined in terms of specific products with specific capabilities, offered at a specific price to a customer. One identifies the value stream, which is a set of

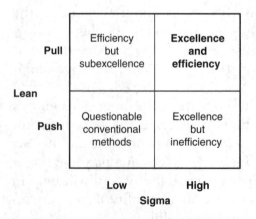

FIGURE 1.2.   Lean–Six Sigma matrix.

specific actions required to create the value. Identifying the value stream for each product or service is a key step in Lean thinking. Typically, the actions in the value stream can be grouped into three categories:

- Actions creating value
- Actions not creating value but unavoidable immediately
- Actions not creating value and avoidable immediately

Having identified the value stream, the wasteful steps are eliminated, and the remaining value-creating steps flow. One opportunity for change is in reducing the batch size. Having a large batch size is such a natural inclination for high-volume producers, to justify a longer setup times. Lean thinking challenges the batch size by focusing on reduction in the setup time. The batch size affects the purchasing quantities, maintenance, setup time, material flow, and quality improvement. One of the measurements of Lean manufacturing is inventory level. Inventory is a good measure of manufacturing woes, because the natural reaction to any problem in manufacturing is to build more or buy more just in case there is a shortage. Suddenly, the cost of carrying the inventory starts eating away the profits.

Managing inventory is like managing a material river, where the volume of flow is dependent on the length, depth, and width of the river. Similarly, in a manufacturing operation, inventory level is dependent on the number of process steps (length), the unique part counts in designs (width), and the organizational policies (depth). When one starts reducing the inventory level, the rocks or problems start to appear. Continual improvement at a dramatic rate is a critical part of Lean thinking for sustaining Lean operations.

With parts manufactured with Lean thinking and Six Sigma designs, the manufacturing or service focuses on the following considerations:

1. Documentation
   - Product or service design documentation
   - Process instructions

- Verification and test procedures
- Repair, rework, redo instructions
- Handling of nonconforming material or unacceptable service

Processes of known capability

3. Simple and shortest process flow
4. Process reproducibility
5. Organizational policies

Products or services designed for Six Sigma built with Lean thinking will be built in a pull system and with virtual perfection. Ultimately, an organization's goal is to produce the highest quality at the lowest cost and with minimum waste. In such an environment, the endless improvement is realized using *kaizen* (continual) as well as *kaikaku* (dramatic) methods.

Michael George, in his book *Lean Six Sigma* (2002), combines Lean and Six Sigma into the total process as speed and quality are intimately linked. Accordingly, Lean and Six Sigma are perfect partners in improving quality and reducing cost and lead time. Actually, a company cannot implement either Six Sigma or Lean without practicing principles of the other. They are inseparable, two sides of a coin.

It has been observed that Six Sigma allows the corporate leadership to establish a vision and cultural change to achieve dramatic improvement. Lean provides visibility to the improvement, at least in the manufacturing operations. Implementation of Six Sigma challenges one to be innovative in all aspects of business, while Lean challenges one to be efficient. Figure 1.3 identifies various aspects of Lean and Six Sigma. Since Six Sigma and Lean complement each other, the combination of Lean and Six Sigma can build the momentum for speedy improvement.

ROLLS-ROYCE TO WAL-MART

In the post–Bill Smith era, Mikel Harry and Richard Schroeder have been recognized as Six Sigma gurus. Although they impro-

SIX SIGMA	LEAN
Typically driven by leadership.	Typically driven by middle management.
Provides a clear focus and target.	Supports the Six Sigma focus and target.
Requires dramatic improvement through innovation.	Results in continual reduction in cycle time.
Requires passionate and inspirational commitment to achieve perfection from CEO.	Requires personal commitment to challenge current processes.
Affects all aspects of business products and processes.	Mainly affects processes for speed.
Driven for excellence.	Driven for efficiency.
Difficult to implement and benefit from Six Sigma in a localized area.	Can be implemented locally in an operation.
Many tools and DMAIC in the tool box.	Fewer tools in the tool box.
Can be the DNA, culture, philosophy, thinking, and standard of excellence of a company.	Can be the philosophy and thinking of a company.
Six Sigma can be achieved without Lean.	Lean requires Six Sigma tools.

FIGURE 1.3.   Six Sigma and Lean comparison.

vised Motorola's Six Sigma approach, Harry and Schroeder must be credited for institutionalizing Six Sigma. They packaged a loosely defined and evolving Six Sigma methodology into the well-defined Breakthrough management strategy, which has revolutionized the improvement process worldwide. Their classic (and now, for some, controversial) work has led to a new Six Sigma industry that has been championing Six Sigma a lot more than the corporations they serve. Such standardization and mass implementation includes the training and qualification of large number of Black Belts (an expensive proposition), Green Belts, and Brown, Yellow, or even White Belts.

Due to their emphasis on standardization of practitioner qualifications, Six Sigma training became a critical component of any Six Sigma initiative. The Black Belt training typically consists of an intensive 4- to 5-week course covering strategic intent, methodology, measurements, and tools. It appears that what one would learn over many years at work could now be learned in 4 to 6 weeks through Six Sigma Belt training. It has been observed that the most significant or obvious skills a Black Belt retains is in Design of Experiment, a statistical method for understanding the effects of many variables, as part of the DMAIC methodology. Such minimal retention of the skill set led ASQ to launch the Six Sigma Black Belt certification test, which tests for the Body of Knowledge, given some hands-on project experience.

Regarding the existing confusion about the Six Sigma methodology and price erosion in the Six Sigma industry, Harry stated in an interview published in *Quality Digest* (February 2004), "I was the Rolls-Royce of Six Sigma. Now my goal is to become the Wal-Mart of Six Sigma, delivered at the speed of FedEx with the quality of Toyota." Without further explanation, this may add to further confusion, regarding whether someone can implement Six Sigma overnight, like a FedEx delivery, or with about 100 defects per unit, like the automotive industry.

One thing is for sure—changes in Six Sigma implementation are expected due to evolution of the methodology based on lessons learned, as well as competitive pressures to make Six Sigma methodology more efficient.

BACK TO MOTOROLA

Motorola invented Six Sigma in 1986 and released it on January 1, 1987. Since then there has been a sense of satisfaction at Motorola that its initiative has helped not only Motorola but many other companies as well. Bob Galvin, who actually saw the potential of Six Sigma in its early development, and was the first CEO to implement it, has said that even though GE, Caterpillar, Sears, and Honeywell appear to be championing Six Sigma passionately, they always recognize Motorola's contributions. There is a sense of pride in inventing

Six Sigma on one hand, and sense of dissatisfaction at being identified as a struggling company on the other. Experts at Motorola have been working on improving the Six Sigma methodology. It has been recognized as an overall business improvement method rather than just a measure of goodness or a methodology for defect reduction.

Matt Barney and Tom McCarty, in their recent book *The New Six Sigma* (2003), list key leadership principles of the New Six Sigma. These key leadership principles are to Align, Mobilize, Accelerate, and Govern. *Align* consists of linking customer expectations to the core business strategy and processes. *Mobilize* consists of empowering people, organizing teams with clear objectives, and training on demand. *Accelerate* requires learning through project-based training, and driving projects to timely results. *Govern* principle consists of implementing superior execution and managing through scorecards. Hierarchical metrics provide the big picture for executives and enable process improvement by doing when needed. As the New Six Sigma has evolved from its predecessor, it utilizes time-tested best practices such as Voice of Customer, Balanced Scorecard, and Accelerated Improvement.

No matter whether the old or new Six Sigma, innovation has been identified a critical component for the realization of breakthrough solutions. One way to differentiate between continuous improvement and Six Sigma is that whereas the objective in implementing the continuous improvement process is incremental improvement, implementing Six Sigma implies a dramatic improvement. In the Six Sigma methodology, the goal is to achieve a significant rate of improvement. New Six Sigma enforces that by accelerating improvement. Given that every corporation is improving its processes, the competitive advantage can be gained by accelerating improvement.

SUSTAINING SIX SIGMA

At the launch of a Six Sigma initiative that will set expectations for superior results, require dramatic improvement, demand major changes, transform corporate culture, and embody the corporate philosophy, strong leadership is required. Strong leadership creates a target for employees, inspires commitment and action from employees, earns the respect of employees, and recognizes employees' successes. Employees must be able to experience the leadership's passion for practicing Six Sigma and realizing measurable results.

One of the main tenets of Six Sigma is measuring what you value. Corporations have been initiating Six Sigma without a measure of their progress in sigma levels. Instead, they have been monitoring growth and profitability. Six Sigma provides the necessary tools and builds the momentum to achieve benefits fast, and corporations realize those benefits. Without a measurement of progress toward sigma levels, one is accepting partial results because they are better than what could be achieved without Six Sigma. Interestingly, corporations that have been implementing the Six Sigma initiative appear to have no drive to establish the corporate sigma level. This is understandable, because it is rather easier to simply pick the low-hanging fruit and not worry about the sigma level. However, to go beyond the obvious, the leadership must decide to take the journey beyond the first few years of Six Sigma. Otherwise, the Six Sigma initiative, is bound to change into another initiative, because the drive for improvement would no longer be Six Sigma, just the elements of it. An integrated, well-aligned, directed, and monitored Six Sigma initiative will extend the life of the improvement and realize sustained results, irrespective of any leadership changes.

SIX SIGMA ON SIX SIGMA

Six Sigma has moved from the Rolls-Royce phase to the Wal-Mart phase, that is, from the evolution to the maturity phase. Still, lots more large corporations have been practicing Six Sigma than smaller companies, due to lack of knowledge and the costs associated with the training and consulting. The Body of Knowledge for Six Sigma certification is extensive and requires a personal desire to practice the Six Sigma tools. The Black Belt training, which used to cost as much as about

$50,000 per person, has to come down even below the current levels of about $15,000 per person. The critical skills required for a Six Sigma professional to be effective are superior leadership, statistical thinking, and a learning attitude, rather than just the expensive training.

Application of Six Sigma is still growing in diversified industries and countries all over the world. Six Sigma failures will become more apparent and publicized because of the frustrations and mobility of practitioners. Corporations will need to apply Six Sigma principles to the Six Sigma methodology. In other words, one must look to optimizing implementation of the Six Sigma methodology to improve return on investment in the Six Sigma initiative.

Six Sigma has been in existence for over 15 years. Breakthrough changes in the Six Sigma process include the use of software tools for statistical analysis instead of spending time doing computations, allowing more time for interpretation and experimentation. Other developments include the Breakthrough methodology, the Lean and Six Sigma combination, and the Six Sigma Business Scorecard. Attempts to optimize the Six Sigma process have been made in this book by applying the Theory of Constraints to the Six Sigma process; however, more work will have to be done to upgrade the Six Sigma methodology.

SIX SIGMA CHALLENGES

Leaders are convinced that Six Sigma is the mantra for performance improvement. Demand for implementation of Six Sigma is still growing. However, the number of successes is not matching the number of implementations. There are more successes published at the project level, where DMAIC has been implemented, than at the corporate level, implying that corporations are not getting the return on their investment in Six Sigma or are not realizing benefits of the proportions of those at GE, Motorola, Allied Signal, Raytheon, and Bank of America. The current challenges in implementing the Six Sigma methodology at corporations include the following:

- Too much focus on the Black Belt and Green Belt training without identifying projects for improvement after the training
- Lack of understanding of the strategic intent of Six Sigma
- More local implementation than companywide implementation
- Too many books and too much talk about Six Sigma
- Lack of innovation in problem solving, no real breakthroughs
- Cost of Six Sigma training and consulting
- Too much focus on projects (more than 10,000 in a company) instead of processes
- Exploitation by the Six Sigma industry of corporations because of the demand and supply differential
- Lack of support for small businesses in implementation of Six Sigma
- Fragmentation of various corporate initiatives
- Too often considered a quality improvement initiative rather than a business imperative

CHAPTER

TWO

BALANCED
APPROACH TO
PLANNING

A balanced approach to Six Sigma is about getting from desire to results, realizing the Six Sigma initiative, and being recognized for successfully implementing Six Sigma. A balanced plan is more than a financial focus on short-term results. It is a balance between that real need and the investment to implement Six Sigma, create long-term value, and create wealth. Focusing on a number of key areas and understanding what enables work on those areas to succeed makes a balanced plan.

A balanced plan is about Six Sigma Strategy, Leadership, Resources, and Execution. It is about setting the right goals, making them real so that employees know what to do, why they are taking the actions they are taking, how those actions contribute to the total picture, and what the end game looks like. A balanced plan for Six Sigma focuses on the strategic intent that will move the company forward, understanding the critical requirements in each area and the key indicators of progress. Managing a balanced plan is based on utilization of facts and data, measurement, reaction, and taking action on the performance results.

The balanced approach to Six Sigma is about interlinking analysis of opportunity, development of mission and vision, establishment of stretch goals and determination of strategies for achieving them, implementation, execution, and improvement. It is also about the supportive elements required to enable employees to engage in those items and actions: role of leadership, alignment, organization development, and management of change.

A balanced plan should create transformational change. It is not a project, an annual event resulting in a book that goes on the shelf until next year, or the passion of a single group of leaders. It is a driver that is present throughout all areas and aspects of the business. It is big and aims to move the company to a new place. The impacts of technology, information, competitive swiftness, financial pressures, and increasing stakeholder demands can be some of the causes driving the need for transformational change.

Leaders pride themselves on their company mission and vision statements. They are posted in the halls and the boardroom, listed in the annual reports and on the employee badges. Leaders declare organizational victory when employees can recite the mission and vision statements. More often than not, this is only memorization and nothing more. Some companies use the terms interchangeably. Statements are finalized, posted, and an expectation is fulfilled.

What would happen if all the posted mission and vision statements in companies were secretly switched during the night? Would employees take different actions the next day? Would customers react differently? Would the organization change priorities and projects? Would anyone even notice the change?

Therefore, the company mission and vision statements around Six Sigma should be living documents that guide the organization, providing purpose, direction, and intent. Leaders must work to support bringing these statements to life. They must create relevance and content for each employee and stakeholder.

The objective is to have a guidepost to help employees know exactly how their individual roles contribute to the greater whole of the organization.

CONVERGENCE OF THEORIES

In the mid- to late 1990s, several significant lines of thought developed and emerged in management literature. Among these were Intellectual Capital/Knowledge Management, the Balanced Scorecard, and the Revitalization of Six Sigma. Practitioners have pushed these theoretical management approaches forward so there are now robust cycles of learning and actual results associated with them. In the past few years, practitioners have begun combining some of these tools and principles to support even more powerful balanced approaches to creating transformational business change.

Examples of three veins of management theory are highlighted in the following paragraphs: Intellectual Capital, Knowledge Management, and Intangible Asset Growth by Leif Edvinsson and Michael S. Malone, from their book *Intellectual Capital: Realizing Your Company's True Value by Finding Its Hidden Brainpower* (1997); balancing short-term gain with long-term value creation by Robert S. Kaplan and David P. Norton, from their book *The Balanced Scorecard: Translating Strategy into Action* (1996); and the *Revitalization of Six Sigma* by firms like GE and Honeywell and the Convergence of Methodologies by the author, from his book *Six Sigma Business Scorecard* (2003).

INTELLECTUAL CAPITAL, KNOWLEDGE MANAGEMENT, AND INTANGIBLE ASSET GROWTH

The foundation of this work was the groundbreaking theory that the true long-term value of a company was made up of both tangible and intangible assets. Each could be measured, grown, and quantifiably valued. Traditional financial focus was balanced with focus around Customer, Process, Renewal and Development, and Human Empowerment. The value potential through management of intangible assets often surpassed the value of tangible assets.

It has been recognized that the true value of a company lies in its ability to create sustainable value by pursuing its vision and its resulting strategy and maximizing certain *success factors*. These success factors could, *in turn*, be grouped into four distinct areas of *focus*. These factors are *Financial, Customer, Process,* and *Renewal and Development.* These factors can be maximized by the empowered human resources.

Within each of these four areas of focus, businesses could identify numerous key indicators to measure performance. This provided a much more balanced approach to planning. Edvinsson and Malone presented a methodology for determining what intangibles should be prioritized to build long-term value.

BALANCING SHORT-TERM GAIN WITH LONG-TERM VALUE CREATION

Kaplan and Norton's landmark position presented a new business operating environment that strongly acknowledged influences of the Internet and the Information Age. They presented the position that succeeding in this new environment created the need for new skills and abilities and the development of intangible assets as the economy shifted. A structured and balanced approach to building long-term value is now needed to balance the traditional financial focus.

Accordingly, a breakthrough in performance requires changes in measurement and the accompanying management systems. The objectives and measures of the Balanced Scorecard are more than just a somewhat ad hoc collection of financial and nonfinancial performance measures; they are *Financial, Customer, Learning and Growth,* and *Internal Business Practices.* These measures must support the corporate vision and strategy.

This model integrated the traditional financial objectives for short-term gains with the investment in intangible growth for long-term value growth. It was presented as both a management system and a measurement tool.

REVITALIZATION OF SIX SIGMA

The commonality between the Edvinsson and Malone and the Kaplan and Norton methodologies begin with using Customer Voice and Requirements as a driver for all other actions. This was also the driver in Six Sigma. Although Six Sigma was widely used earlier for driving quality improvements, it gained new life in the mid- to late 1990s as a tool for driving transformational change. With the evolution of the DIGMA structure at GE came more structured methodologies for driving strategies and objectives down into the operating fabric of the business. Companies such as GE and Honeywell began to use the system structure to drive transformational change across the entire company.

Within the DIGMA approach, and its numerous derivations, are the base elements of *Customer, Internal Processes and Practices,* and *Institutional Learning* supported by strategy and objectives.

Six Sigma therefore provides a tool for bringing life to missions and visions.

Note the common elements of each of these theories and methodologies, as shown in Figure 2.1. Each was a response to the changing business landscape and emerging needs. Each presents shifting attention from exclusively focusing on short-term results to a balanced approach of pursuing short-term

PARAMETER	INTELLECTUAL CAPITAL/ KNOWLEDGE MANAGEMENT	BALANCED SCORECARD	SIX SIGMA
Leadership			■
Financial	■	■	■
Customer	■	■	■
Internal operations	■	■	■
Learning and growth	■	■	■
Vision and strategy	■	■	■
Process-level objectives			■
Measurement system	■	■	■
Rate of improvement			■

FIGURE 2.1.   Comparative analysis of various methodologies.

results *and* building long-term value and creating wealth. Each provides a measurement system to quantify the development of intangibles.

Six Sigma Business Scorecard.   This management tool combines the power of a balanced management approach between intangibles and tangibles, from Intellectual Capital and Balanced Scorecard, with the analytical power of the Six Sigma methodology. It produces a balanced measurement structure to drive improvements across all levels of large organizations, resulting in a single measurement value that is a indicator of overall company health.

Key components of the Six Sigma Business Scorecard model include *Leadership and Inspiration, Management and Improvement, Sales and Distribution, Services and Growth, Employees and Innovation, Operational Execution,* and *Purchasing and Supplier Management,* supported by an interlinked circle of Improvement, Revenue, Innovation, and Cost.

This methodology brings the capabilities to drive change down further into the operational and goal level. It brings forward the balanced focus between tangibles and intangibles, short- and long-term growth.

As the times and the business landscape have evolved, the leading-edge methodologies have begun to converge and build upon one another. Leaders and practitioners are matching appropriate tools to their needs while maintaining the newer balanced focus. The basic elements of each system have proven to be solid and valuable in business planning.

This chapter focuses on creating a strong thread between the planning components and the associated leadership responsibilities, organizational development, and change management aspects of creating a balanced plan and carrying it through to achieve the desired results.

LAUNCHING A SIX SIGMA INITIATIVE

Capitalizing on the right Six Sigma opportunity can be profitably rewarding. Before committing to the Six Sigma initiative, one must evaluate the opportunity through a structured, fact-based, objective, and balanced view. Analyzing the opportunity is the first step in building a plan for successful implementation of Six Sigma. A small team of senior leaders, with a vested interest in the business and in the Six Sigma initiative, should do the analysis. The group makeup should represent interdisciplinary functions and interests. The team should focus on the effects on both the tangibles (financial numbers, hard assets) and the intangibles (abilities, soft assets).

To be complete, the team should evaluate the opportunity and the business side by side in a *who, why, what, when, how,* and *where* approach.

THE WHO—UNDERSTANDING THE BUSINESS

Reaffirm and Articulate Your Mission. Articulating and succinctly defining your business is difficult but also fundamentally necessary. Clarity and brevity are required. Numerous management books cite how few companies can truly articulate the business they are in and the power of defining *the business*. A well-articulated business definition is like the soul of the company. It has an empowering influence that supports action. When articulation is achieved, the definition becomes the driving basis behind the company's mission and all strategic decisions.

Both internal and external stakeholders will respond to the business definition. It broadcasts the message of a company's essence. It begins to identify and organize a business's core focus, priorities, actions, and capabilities.

Consider the example of Coca-Cola. Is it a cola business, a soft drink business, a beverage business, or some other business? Numerous references present evidence that suggests that Coke has defined its business success as share of *liquids* consumed, not share of cola, or soft drinks, or beverages. The importance of the business definition can open up new potentials for the company's objectives.

Opportunities, challenges, or desired changes may require you to revisit how you define your business. Investigation of new opportunities may identify approaches or values that cause

you to broaden, narrow, or otherwise alter the business definition. It is imperative that your employees, customers, and other
stakeholders clearly understand who the company is and what
business you are in!

Define What You Want to Be Known For.   A corollary to the
business definition is to articulate what you want to be known
for. This addition to your business definition begins both to
shape your company vision and expand your possibilities for
the future. Mission statements are broad and can be inter-
preted differently by different people. More specificity begins
to create a shared mindset. Two or three carefully worded sen-
tences can be used powerfully.

Extrapolating an extreme statement from the preceding
example, the company may want to be known for *providing the
best-tasting and/or highest-quality consumable liquids.* This is
understood to currently include cola, flavored colas, and water.
Might it advance into other liquids, such as dairy products,
infant formulas, or wine and alcohol-based products? What
about liquids consumed by animals? What about medical liquid and fluid products?

All of these fit into the "share of liquids consumed" opportunity. The exercise of articulating what you want to be known
for can both prevent misinterpretations and highlight new possibilities.

A few simple definitions can provide a framework to accelerate strategy development. Can all functions in the company
align behind those definitions? Can all areas contribute and
affect performance in the defined areas? Will establishing this
image with your stakeholders, especially customers and
employees, provide advantages in the marketplace? Two or
three articulated statements can serve as guideposts to the
organization.

Use the Customer Voice.   Without customers there is no
business. Understanding the requirements, desires, and opportunities of customers is core to all business methodologies. The
customer voice is a critical component of the Six Sigma Business
Scorecard, Performance Excellence, Malcolm Baldrige, Six
Sigma, Balanced Scorecard, Knowledge Management, Total

Quality Management (TQM), and other major transformational methodologies. In these systems, the customer voice positions as the driver of the entire process, the hub of the system, one of a limited number of prioritized drivers, or the definer of success. Consistently, the customer voice serves an important role in the company opportunity.

Bring customer knowledge and input to your analysis and planning. Tapping into those employees who actually have contact with your customers can be beneficial. They may have some enlightening inputs on opportunities to provide more value to the customer. They will also have insights on how the company product or service is actually received and perceived by your customers. Ask customers directly. They will usually be open in sharing their view of the competitive marketplace, a key contribution to your planning process. Allow customers to share their strategic plans and objectives with you. Look for opportunities for leverage, collaboration, and mutual wins.

In the end, the customers will determine your company success, and they will vote with their currency and continued business demand. Bring their inputs into the front end of activities to increase your opportunity for success.

THE WHY—UNDERSTANDING THE DRIVER OF OPPORTUNITY

Be clear about why you are looking to develop a new plan. Is there a crisis, a unique opportunity, or both? The words *challenge* and *opportunity* may be used to describe the same situation. *Opportunity* is used here to describe both perspectives, believing optimistically that even the worst influence can be turned to a positive direction.

Response to drivers may require defensive actions or be offensively leveraged as opportunities. Either way, it is important to understand what is behind the need to evaluate opportunity. Clearly defining the driver will allow a structured approach to creating a solid plan for response. Evaluation and analysis must be a balance of fact-based and market intelligence using data and best information. Decisions should not be based on subjective guesses or emotional inputs.

Some drivers are based on internal influences, such as the need to improve performance, create new products, or increase the rate of growth. Other drivers can be categorized as external or disruptive influences, such as customer shifts, competition maneuvering, or technology surprises. Response to each type may be approached differently.

Internally Initiated Drivers

Business Survival.   If a business is faced with survival, there is motivation to make drastic moves and monumental decisions. This can be beneficial in aligning employees on critical requirements. Special care should be taken to avoid compromising needed core competences and capabilities in the process of defining renewal. Short-term objectives must not completely come from shedding resources and tightening budgets.

ACTION: Evaluate needed results in several time frames: immediate, interim, and long term.

Growth Potential.   Opportunistic growth may be focused around new products, lateral or vertical market integration, distribution methods, geographic expansion, or acquisitions or mergers. This may also be unrealized growth from the existing overall strategic plan.

ACTION: Evaluate the balance between various elements contributing to growth potential.

Advancing Long-Term Value Creation.   Development of a critical capability, competency or asset may provide a competitive advance. Building intangible value may have a strong positive impact on stock values.

ACTION: Identify areas of strength and areas of potential.

Critical Performance Improvement.   New methodologies, new capabilities, or new talent may present opportunities to improve internal performance. Operating performance improvement may be necessary to remain competitive. Consolidation, moving to shared services, or even outsourcing may be drivers.

ACTION: Reaffirm company values, determine desired strategic capabilities and competences, and evaluate current trends and results.

Building Additional Capability.   Demands or opportunities may be outside of current company boundaries of operation or capability.

ACTION: Evaluate potential benefits of acquisition, mergers, partnerships, alliances, subcontracting, new talent, and investment in systems.

Annual Planning Cycle.   Missions and visions are about company essence and the longer-term future. They should not change significantly year over year unless there is a major disruptive influence. Some business cultures set an expectation that missions and visions must be updated annually and significantly different. Even with the rapid pace of business in today's information-based world, radical shifts can confuse and distract organizations so that forward momentum is lost. Worst of all, some company cultures change just for the sake of change only! Better to have a strong, future-creating mission and vision, supported by strategies and goals that can be fine-tuned to address the changing business landscape.

ACTION: Reaffirm and evaluate against any new realities.

Disruption-Initiated Drivers.   Disruptive business influences do occur. Good plans anticipate surprises and work through scenarios of response to possible disruptive influences. So your current plan may already address this type of driver.

Competitive Threat.   Unexpected challenge or move into your market, product, service, or operational space by another entity.

ACTION: Increase competitive intelligence knowledge for analysis.

Regulatory or Governmental Requirements.   Rules, laws, guidelines, or requirements for doing business change.

ACTION: Utilize associations, universities, and other institutions to understand implications. Evaluate the benefit of joining alliances to become a larger player in the landscape.

Customer-Driven Opportunity.   Customer desires or possibilities increase in importance to become necessities or absolutes.

ACTION: Increase customer intelligence. Work directly with the customer. Tap into information from internal employees with direct contact to customers.

Technology Advancement.   The convergence of technology and the rapid deployment of applications can create unanticipated opportunities or challenges. An example is the far-reaching, and still evolving, effects of nanotechnology. Advances such as this change the very basis of established business relationships and future potentials. Tracking developments and the implications to your business may help identify potential opportunities before they hit.

ACTION: Ally with research organizations and universities to gain early warning of advances and understanding of effects. Utilize trade and professional associations that monitor technology trends. Participate in and influence standard-setting boards.

Anticipation-Initiated Drivers.   Drivers may be seen on the near horizon but are not yet in complete view or affecting the business. Experienced leaders learn to read the signs and connect the dots, responding with preemptive moves to advance their business activities.

CEO-Created Crisis.   CEOs may create the image of a crisis to launch activities or motivate the organization. Although the intent is usually noble, this approach can be risky. The explanations of why are often weak or even made up. People see through this and become uncomfortable with the plan, disengaged from the required actions, and even rebellious. Therefore, if a created crisis must be launched, the CEO must dedicate significantly more attention to creating a unified direction, providing justification for the goals and personal involvement in execution management.

A clear understanding of *why* to address an opportunity allows:

- More complete analysis
- More robust definition of needed direction
- Realistic communication of need, leading to employee buy-in

THE WHAT AND WHEN—WHAT DO YOU WANT OR NEED TO ACCOMPLISH?

Answering this question defines your endgame, intent, and vision. This is ultimately where you want the organization to focus. It is the win, competitive leadership position, achieved advancement, market dominance, technological invention, or financial improvement that your opportunity is all about. Because the driver is important, the endgame should be a significant achievement, an aggressive reach, and a transformation! It is also positioned out in the future, not the near term, describing a future state. Articulating the *what* and *when* begins to bring life to your vision.

Start with Reality.   There is no way to avoid it. Start with the reality of where you actually are. Use facts and actual data to create an honest and realistic report of your starting point. What is your base? What are the major components of your base? Which ones are strong and where are the weaknesses? How did you get where you are? It is important to have an understanding of how you arrived at the current position. Objective review can identify learnings that will assist with evaluating the current opportunity.

Use only good data. Create a complete picture. You may have to look for some components of the picture, but, with some work, a factual picture will develop. A senior manager who facilitates performance improvement refers to this as "liberating the data," because it is always there. You will improve your own *business intelligence.*

Competitive intelligence must also be part of the evaluation. Where do you stand in relationship to your competitor and anticipated competitors? You should also identify who the best of the best are and where they are. Where is your business positioned in the largest possible view of the landscape?

Defining your actual starting point is required to anchor all other activities and provide the needed foundation to create a solid plan.

Describe the Benefits.   The structure of a high-level business case should be developed. A list of anticipated benefits should be articulated. What is to be gained? More detail will be developed as the process moves forward.

Evaluate the Alternatives.   Determining what you want or need to accomplish should include evaluation of various alternatives. You can start by how you position yourself to view the opportunity or challenge. It can be powerful to look at the opportunity or challenge from both a positive and negative side. Can a critical challenge be viewed through an opportunity lens? Can an opportunity objective be acted upon more completely when described in terms of a pending challenge? Forcing evaluation from both sides of the situation can identify alternative endgames as well as begin the formulation of how the endgame is defined.

Management tools like scenario planning, benchmarking, breakthrough, and workout sessions help leaders identify alternatives through proven structured methodologies. The benefits of *what-if* exercises are built up into robust evaluations. These must be grounded in fact and supported by data to be realistic. Why are you doing this instead of another approach or something else? Can you gain insight or learning from prior efforts or benchmarking studies?

Alternative evaluations will be more powerful and productive if the participants represent diversified functions and interests. This is an initial point where diversified input and a balanced approach create a stronger plan.

Make It Big.   You are creating a sense of purpose and direction. Make it big. Set the stage for transformation of the organization. Set an aggressive vision. Set the expectation for significant change. An associate who previously worked at NASA says, "If you absolutely must make the moon, aim for Mars." His message is to think bigger than what you absolutely must achieve.

Be Specific to Be Inspiring.   Provide a clear definition of the endgame and define the win! Put words around what may seem impossible. Leaders throughout the company must be able to base their independent actions on understanding the vision. More clarity enables higher alignment of their actions.

In addition to your objective, provide supportive details. Include high-level elements such as the organization or group, timing, level or position, measurement, recognition, or area. The objective may become the tag line, but the supportive details will be understood by the organization.

A famous example is John Kennedy's vision of putting an American *on the moon.* In the middle of highly turbulent times and upheaval (think disruptive influences and business crisis), Kennedy communicated a vision that united a nation into action. Think of the thousands of people who applied their energies and the billions of actions that aligned behind his charge to make the vision a reality. His single message was to improve America's technical capabilities by invigorating the space program and all the ancillary benefits it would bring. In the middle of the global arms race, he crafted a vision that supported peace while increasing beneficial defensive/offensive capabilities.

Kennedy's vision was articulated as follows: "I believe [*personal investment*] that this nation [*engaged everyone, not just NASA or a single organization*] should commit [*the charge for action*] itself to achieving the goal before this decade is out [*time specific and aggressive*] of landing a man on the moon [*aggressive goal defined*] and returning him safely to earth [*definition of success*]." Kennedy's vision included specificity of goal, definition of success, and required time frame. It was a single engaging sentence that inspired the achievement of what was then thought to be impossible. Its effects are far reaching, still remembered today in the quotes, "The Eagle has landed" and "One small step for man, one giant leap for mankind." These are inspiring reminders of actual achievements originally thought to be impossible. The achievement was a significant reach, and the results continue to benefit each of us today.

Define Your Six Sigma Vision.   Articulate the endgame. Be bold. Don't just think outside of the box—blow it up! Stretch the present to create a vivid image of the future. Be clear about the direction, and define the endgame in ways that are easily understood by all levels. Show personal investment and leadership; use phrases like *I believe* and *I will*. Define the future state, and give highlights of *why* you are going to lead to this new place. Be clear on the advantages to be gained.

A company's Six Sigma vision provides a future-oriented outlook for an organization. Superior companies establish their vision irrespective of the Six Sigma initiative. When a company commits to a Six Sigma initiative, supportive homework must be done before launching the initiative. Creating a vision for a Six Sigma initiative requires a complete understanding of the Six Sigma methodology, its intent, and its benefits. Without clearly understanding its benefits, the corporate vision could miss its sense of direction. For example, one company understands Six Sigma as a DMAIC, and therefore, a five-step methodology. For another company, Six Sigma has been understood as a strategy to dramatically improve business. Management leaders continue to evolve the application of Six Sigma and create new benefits. Is it a strategy, a methodology, or an enabler? GE sees Six Sigma as its DNA, Motorola sees it as a cultural thing, and Honeywell sees it as a standard of excellence. Six Sigma, which was born as an approach to accelerating improvement, promoting employee teamwork, and achieving total customer satisfaction, has been analyzed to its limits and institutionalized at many corporations. Six Sigma was initially criticized for being unrealistic due to the associated 3.4 defects per million opportunities. Yet today, Six Sigma has become a strategy for improving corporate performance through a culture of continual reengineering and structured self-assessment, a methodology for dramatic improvement and employee innovation used to produce the best products and services at the highest profit.

With this understanding of Six Sigma, the corporate vision must look beyond the number, 3.4 defects per million opportunities. Instead, the vision must incorporate the philosophy, its methodology, and the expected outcome. The vision must not be a glamorous claim to be the best in the world; instead, the vision must be a statement that relates to the desired cultural climate and new levels of corporate performance. The vision must provide the foundation for unifying efforts around both the needs of the business and the needs of the customer. At Motorola, it was a combination of Vision, Beliefs, Goals, and Initiatives.

THE HOW—IDENTIFYING CAPABILITIES AND RESOURCES

Capabilities can be viewed as the organization's ability to facilitate the achievement of the company mission and vision. They are the combined abilities, skills, knowledge, and expertise across the total organization. Capabilities are a combination of what people are able to do, what they know, and how they think. Without the right tools, you may not be able to build what is in your dream.

Ulrich and Smallwood have documented the effects of capabilities and competences on business value and growth in *Why the Bottom Line Isn't* (2003). They present how organizational capabilities represent the ability of an company to use collective resources to get things done and operate in ways that lead to accomplishment. Their model of capabilities include *shared mindset, talent, speed, learning and knowledge management, accountability, collaboration,* and *quality of leadership.*

Ulrich and Smallwood note that to be successful, the *capabilities* support the core *competences* in the technical or functional requirements of *product innovation, operating efficiency, customer intimacy, distribution,* and *technology.* A company strategy should focus on one of these competency areas but also contain elements from several others to create balance.

Use this list of capabilities and competences as a starting point for an initial check against your defined vision. You may find strong advantages in some areas and weaknesses or omissions in others. If a critical area is determined to be weak, evaluate what can be obtained. There are four alternatives for action.

Buy. Can you hire, purchase, acquire, or integrate from somewhere else the knowledge, abilities, or skills needed to complete your vision?

Borrow. Can you contract, retain, lease, partner, or rent the knowledge, abilities, or skills needed to complete your vision?

Develop. Can you grow, train, learn, create an alliance, or accelerate the knowledge, abilities, or skills needed to complete your vision?

Change. If you do not have the knowledge, abilities, or skills needed and cannot buy, borrow, or development them in the anticipated time frame, you must realistically change your vision and plan.

The strategy for how to proceed must follow your capabilities. Development of core competences will focus activities and accelerate actions.

THE WHERE—ESTABLISHING FOCUS

With direction and purpose established, it is now appropriate to define the concept of where to focus the plan. *Focus* means exactly that, attention on only a few things. Real-life practitioners consistently highlight the need for a few simple areas of focus at a time. Organizations have a difficult time spreading efforts and attention across more than three to five objectives.

The Four Critical Areas. It is important to integrate the four areas as you work through the critical areas of focus. According to George Labovitz and Victor Rosansky in their book *The Power of Alignment* (1997), the four areas are as follows:

Voice of the Customer. Customer requirements change over time, and the organization's customer intelligence must keep up.

Gathering actionable data from customers should be through both formal and informal channels. It should also be openly confirmed with the customer to build trust and partnership.

Voice of the Business. Internal objectives and requirements must be acknowledged and included in the plans. These objectives and requirements fund or provide support for growth and improvement. As the plan must be focused on the strategies of the company, the voice of the business must be solidly integrated into the plan.

Voice of the Employee. Employees must have the capability to execute the plan objectives. Their understanding and buy-in to the plan affects their ability to carry it out. A balanced plan must acknowledge both the current status of employee capabilities and the objective target for new capabilities and cultural climate.

Voice of the Key External Forces. External forces that must be adapted to must be included. Examples may include external regulatory rules, competitive pressures, and technology shifts. Intelligence on trends and pending changes must be evaluated for planning.

Moving from Issues to Critical Focus Areas. Each of these voices will have items that are critical or imperatives and others that are desired or attractive. Distilling the inputs down to a critical few develops the critical must-have items to focus on. Each critical requirement must be measurable, and a target must be definable. Keep the number limited, the definitions simple, and the targets in reach.

It is essential to translate the critical requirements into what must occur in your business. What affect in which area must be achieved? This provides the cause-and-effect relationship between the *voices* and your business activities, as shown in Fig. 2.2. The final column of Fig. 2.2 becomes the foundation for your key measurement indicators.

Establishing this priority sets you into actually creating your plan. Pay attention to creating balance across your business organization and operations to create a plan of linked objectives.

INPUT	ISSUES, AREAS OF FOCUS	DETERMINED TO BE CRITICAL REQUIREMENTS	RESULTING FOCUS AREA OF WORK
Voice of the customer	Customer issues, objectives, opportunities	Critical customer requirements	Critical to quality and customer
Voice of the business	Business issues, objectives, opportunities	Critical business requirements	Critical to process financials
Voice of the employee	Employee issues, objectives, opportunities	Critical employee requirements	Critical to learning and improvement
Voice of the key external environmental forces	Competition, technology and regulatory issues, objectives, opportunities	Critical environmental requirements	Critical to change response

FIGURE 2.2.   Voice of the customer and customer relationship.

Creating the Plan

Conduct Business Opportunity Analysis.   The success of a Six Sigma initiative begins with a good understanding of needs for Six Sigma. The business performance is evaluated along the major business and functional areas: Customer, Leadership, Operations, and Suppliers. The need for Six Sigma is established based on the analysis of a company's performance in areas as identified in the Six Sigma Business Scorecard, such as the following:

- Leadership philosophy and values
- Recent improvement activities
- Employees' state of mind

- Corporate culture
- Financials
- Operations
- Management systems
- Performance levels and trends

Data are collected from the finance, manufacturing, engineering, sales, purchasing, and quality departments. Key managers are interviewed to understand their operations and challenges. A comprehensive report and presentation is generated on the outcome of the data and observations analysis. Certain assumptions are made to estimate the performance where no data are available. Based on the analysis and observations, the report identifies key opportunities for improvement and their potential effect on profitability. A sample of strengths and challenges of an analysis are as follows:

Strengths
- Innovative solutions
- Product quality
- Positive employee attitude
- Proactive customer interaction

Challenges
- Use of old data for the quotation process, leading to variance in actual versus quoted
- Heavy workload and last-minute "firefighting"
- Internal rework in almost every job, including a lot of non-value-adding activities such as inspection
- Delinquency due to nonavailability of parts and inability to track material
- Glitches in software development, leading to project delays
- Proliferation of old designs and poor documentation, leading to waste
- Delay in delivery of material to shop floor after its receipt
- Unacceptable financials

Establish the Baseline.   A balanced plan for improvement must begin with a starting point. Therefore, it is imperative that a factual picture of where you are be established. This should be based on the areas identified for your balanced plan and should be data supported.

Depending on the measurements that represent your business and will drive your plan, data should first be obtained from existing resources, if possible. *Caution:* Do not establish a measurement just for the sake of a measurement; do not let it take on a life of its own. Use what you have. Let financial, customer satisfaction, sales, quality measurement, employee participation, and leadership performance data do the talking.

Figures 2.3 to 2.5 show various performance measurements to understand and identify opportunities for analysis. There may be several more measurements to be analyzed to identify opportunities for improvement. Typically, the following measurements are evaluated in identifying opportunities for analysis:

1. Performance or quality data for operations
2. Customer feedback or relationship information
3. Sales performance
4. Marketing data (market share, market size, etc.)
5. Internal or customer audit reports, if any
6. Employee ideas or suggestions
7. Employee recognition
8. Purchasing and suppliers performance
9. Inventory level and trends
10. Financial reports, including balance sheets for past 3 years
11. Business plan or equivalent document
12. Competitive analysis
13. Monthly or quarterly operations review reports

Besides the preceding information, one must review operations and interview a sample of employees to capture their

Product/Service	Market Size, $ Million	Market Share, %
Custom products	50	10
Standard products	20	90
Services	60	20

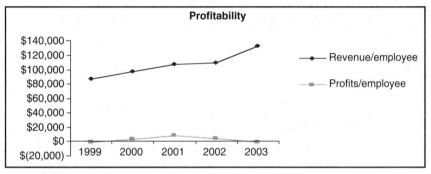

FIGURE 2.3.   Analysis of market share and bottom line.

observations and recommendations for making their company better.

Translate Opportunity and Vision into a Balanced Plan

CREATING A CONSTRUCT.   We have learned over time that in developing a business plan for turnaround or performance improvement for growth and profitability, companies often lack a framework. A strong framework will allow an organization to address the financial, operational, and intangible aspects of the business with a balanced view. A framework pulls leadership, management, and employees in the same direction. In absence of such a framework, companies start with financials, sales projections, and some operational data. However, due to non-performance, the management may find it hard to follow up and get excited about results. It is critical that leadership select a framework that would suit its business and develop its business plan to achieve the desired results. Given the current degree of competition and economic environment, Six Sigma must be a part of the framework. The most critical element of the framework that actually drives improvement is the rate of

48

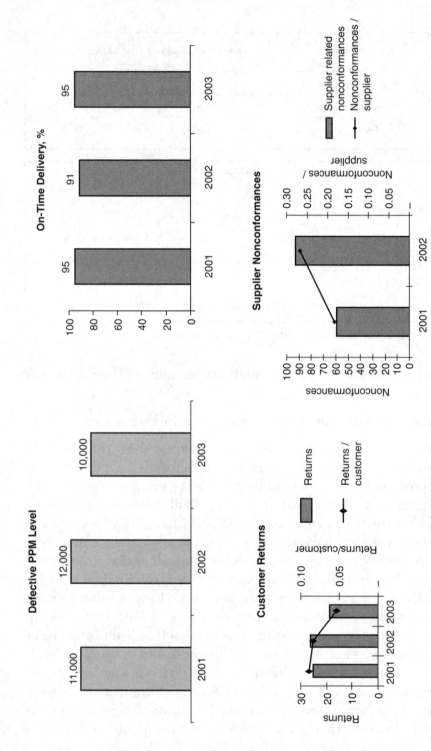

FIGURE 2.4. Analysis of operations performance.

OPPORTUNITY	SAVINGS POTENTIAL, % OF SALES
Increase project efficiency	4
Improve inventory turns	3
Reduce total cost of purchasing	2
Reduce project cycle time	1.5

FIGURE 2.5.   Identifying opportunities for improvement.

improvement. This has become the competitive advantage, as most companies are targeting sustained levels of continual improvement in their performance. The Six Sigma Business Scorecard provides such a framework (Fig. 2.6).

The balanced approach must consider the needs of a corporation through its vision and objectives. The balanced approach must ensure the adequacy of planning, collaboration among various players, and the definition of clear roles and responsibilities.

STRATEGIZING TO ACHIEVE OBJECTIVES AND GOALS.   People tend to support what they have created rather than what has been imposed or forced on them. The executive team establishes key market and customer-driven strategic initiatives, then engages employees in the goals and objectives. The strategic planning begins with training in the consistent planning process. Successful implementation of a strategy for transformation requires prioritizing the elements, determining which are the most critical, and executing the most important elements first. Various corporate initiatives must be critically reviewed for the effects of technology, and their effects on people, operations, finances, communication, and business performance.

In superior-performance organizations, the leadership encourages the establishment of stretch goals that require extra effort but, more important, challenge the status quo. The fear of missing stretch goals must be eradicated by recognizing efforts to achieve stretch goals and the achievement of significant improvement. Sometimes goals take several years to com-

IMPROVEMENT

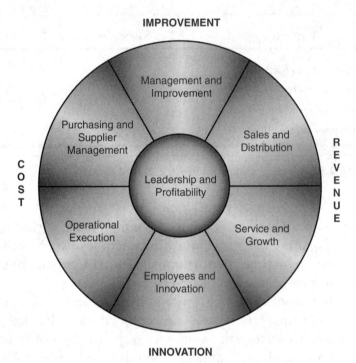

INNOVATION

FIGURE 2.6.   Six Sigma Business Scorecard framework.

plete; in that case, one must plan for quantifiable and measurable intermediate goals to affect behaviors preserving the status quo. Stretch goals are ambitious and highly targeted opportunities for breakthrough improvements in performance.

Establishing procedures for interpreting results with respect to goals and developing an action plan helps the management team to drive improvement. The balanced approach requires that each member of the management team take ownership for improvement in their own area. The manager must drive for better, faster, and cost-effective performance continually.

Initially, one should consider developing a structured approach instead of wide involvement of levels. More levels in the company can be involved as progress is made. The structured and progressive approach mitigates the risk of failure, provides experiential learning, and ensures balance and reality. The structured and balanced approach to the Six Sigma initiative leverages other efforts in achieving the strategic goals.

CRITICAL RESPONSIBILITIES OF THE LEADER FOR SIX SIGMA

Leadership and management roles are defined differently. General Gordon Sullivan and Michael Harper have noted in their book *Hope Is Not a Method* (1996) that *management* has to do with an organization's processes and their execution and its actions, while *leadership* has to do with an organization's purpose and objectives. The CEO establishes the commitment to Six Sigma, while the management ensures its implementation.

The CEO, or key leader, must have an honest conversation with him- or herself. Do the Six Sigma vision, strategic objectives, and goals create enough excitement and energy that the CEO is willing to make the *personal investment* required to lead the plan forward? Is the CEO willing to tie his or her image, success, and compensation to the success of the Six Sigma initiative? If the leader is not intellectually and emotionally engaged in the plan, there are serious questions to answer about the success of the plan.

One can observe that true leaders demonstrate passion for their work. Their actions demonstrate personal investment in attaining customer satisfaction, business profitability, and employee development. Leadership and servitude are in balance and visible through their actions. They openly communicate, demonstrate clarity of intent and mission, establish relevancy of each job to the objective, and build an engaging future. They establish accountability but balance it with investment in capability development. Those who act as leadership role models demonstrate many of the traits needed to lead a plan to success.

Following are key leadership responsibilities. Senior leaders and managers consistently cite this brief list of the CEO behaviors needed to execute a plan.

OWN THE VISION

Take Six Sigma very personally! It has to be the highest priority. Transforming a company using Six Sigma and achieving new levels of performance are not a part-time job or a small project. Make achieving the desired results a consistent priority. No one, at any level, internal or external, should question your vision. Integrate plan reviews and activities into the fabric of your management of the company.

Lead from the Front

Be the face of the Six Sigma vision and objectives. Be visible and vocal! Demonstrate real support for the team's efforts. This includes everything from meeting attendance to resource allocation to participation in reviews. Stay with the program and lead by example. Model the collaborative behaviors required to bridge operational boundaries and establish accountability.

Establish the Direction

Articulate the vision and make it valued. You have the ultimate responsibility to launch the organization on a focused course on the vast sea of possibilities. Others within the organization will develop the tactical actions to support the direction. All levels of the organization must receive a consistent, unwavering message.

Actions, conversations, and messages must communicate your direction and expectations. A vision has been established; push it forward at every opportunity. Build confidence about the future. Don't let people focus on the past.

Align People

Facilitate a unified understanding and forward movement among the organization. Combine and orchestrate various strategies to achieve the desired balanced results. Disband projects or operations that are not in alignment with the direction or activities of the goals. Work and communicate with anyone whose efforts or cooperation is needed to achieve the vision. Support those who demonstrate leadership across lateral boundaries. Create shared risks–and–reward structures. Proportion accountability beyond the current personal and functional boundaries. Focus attention and actions on the end goal.

MOTIVATE AND INSPIRE

Walk the talk! Create a shared mindset among all stakeholders. Provide clarity around objectives. Help people become energized and engaged in activities. Show your excitement and energy. There is a large marketing role to play with both internal and external stakeholders. This is not just focused on employees but also on partners, suppliers, and customers. Acknowledge oncoming change and help people anticipate barriers. Strive to create trust and buy-in so that people will invest themselves. People generally want to have an effect on their work; liberate this desire toward the company's vision.

SET REALISTIC EXPECTATIONS

Achieve aggressive objectives by linking stretch goals. Build hope and instill confidence by setting goals that are a reach but attainable. Unrealistic objectives are deflating and counterproductive and can cause disengagement of employees. Attaining challenging but realistic goals builds momentum and success.

Attainment of a stretch goal is like building a muscle. Each time a repetition is achieved, the muscle grows stronger. It makes the next repetition easier. Runners do not attack a substantial goal, like a marathon, without staged training and achieving ever-increasing levels of success. If they attempted the entire marathon run on a first effort, without achieving their interim goals, they would risk severe muscle damage (resource and capability damage), negative mental challenge and demotivation (desire to disengage), and potential long-term damage (inability to ever achieve the ultimate goal).

The same is true with achieving an aggressive company goal. Linking and achieving stretch goals builds momentum. An aggressive objective will be achieved more quickly with an increasing sequence of linked stretch goals. This builds skills, engagement, and interim results and protects against damage to core capabilities.

ESTABLISH ACCOUNTABILITY

Manage results! This will make or break your efforts to achieve the desired results. Respond and act on both the achievements and the misses of the plan. If a commitment is made, is it accomplished? Communicate positive and negative consequences. Carry through on them. It's that basic. Your behavior in establishing accountability will be mirrored by the entire organization.

Use review meetings and formal reporting for short-term management. Use compensation and recognition for long-term management. Establish a consistent behavior for the operational culture to model. Follow through. Hold the team responsible as well as individual members.

COMMUNICATE

When people understand the *why* of an objective they accept the *what* of activities more quickly. They also apply their energies more freely. Share important information about the business to enable their actions. Use all types of media to communicate regularly with the entire organization on both the objectives and status of the plan. Make sure there are connection points at all levels to increase both vertical and horizontal communication flows. Reach out, question, listen, and obtain information from those executing the plan activities. Communication must be a two-way exchange.

USE THE PLAN—MARKET AND PUBLICIZE

Take every opportunity to increase understanding of objectives and goals. The CEO role has power and respect that must be used to prioritize actions. Facilitate dialogue among and between groups. Observe and listen to what is said. Take action to increase understanding around issues that may arise. If done well, your marketing messages will soon be vocalized throughout the organization.

KEEP THE FOCUS

Tie all organizational activities to the desired results. Manage disruptive influences that can distract and dilute efforts. Hold the team accountable to the agreed-upon actions. Manage potential scope creep. Communicate effectively about required trade-offs if required. Recommunicate priorities.

Likewise, it is important that the CEO not allow other issues or opportunities to distract his or her own attention from the plan activities. A plan can be updated and integrated with new goals, but guard against parallel plans or activity paths that will redeploy resources. Distraction at the CEO level sends damaging and conflicting messages to people. Be consistent in your actions and focus.

STAY INVOLVED

Show up and give people the gift of focused time and attention. Again, actions speak louder than words. If achieving a goal is important, the CEO will remain actively involved until the results are attained. Time commitments may ebb and flow over the plan period, but the CEO's involvement must be present.

Most CEOs, or key leaders, may say they fulfill these responsibilities as a company plan is launched and executed. Yet, those who assist in executing strategic plans consistently call out the preceding items as areas of need. They passionately highlight how more CEO attention is needed for the plan's success.

The CEO, or key leader, has fundamental responsibilities in achieving Six Sigma results. Visible, vocal, pervasive ownership of Six Sigma is a must. Responsibilities extend to establishing behaviors regarding the implementation of Six Sigma. While other leaders, managers, workers, and partners support the CEO, each of these groups looks to the CEO for leadership.

IMPLEMENTING PLANS

Implementing a plan is not a day-to-day activity or taking the next action step. It is the activity of guiding or controlling the overall plan and making any required adjustments to keep it focused on the endgame. It is the act of arranging resources and events at a strategic level. It is refinement of the *construct* that was developed as part of the objectives planning phase.

TOOLS FOR IMPLEMENTING

As the leader begins to implement the plan, the following items should be complete and available.

Balanced Business Plan. This should include articulated strategic objectives, well-defined goals in key areas, business analysis, and a grouping of interlinked project plans. The focus should be well established and clearly defined. The reasoning behind selecting the areas of the balanced plan should be documented.

Project Plan. The interlinked project plans should include the project definitions, business case analysis, scope definitions, and activity plan. Key performance measures must be agreed upon.

Alignment Effects. An outline of the links and relationships between projects should be mapped and described. Key leaders or stakeholders in each focus area should be identified. Any effects on other areas or units should be highlighted and monitored.

Effect of the Customer and Business Voices. The intent of each arm of the plan should be clear. The effects on critical customer, business, employee, and environmental requirements should be documented. During implementation, feedback from each of these voice sources should be used.

Team Information. The leader must be able to delegate the day-to-day execution to key team members. He or she must know the role each team member is assigned. Knowing the responsibilities, capabilities, and expectations of each member is a base requirement. The charter given to the team should also be well documented.

The leader must arrange things and make them fit together, in a networked support manner, so that activities support each other. All resources and activities must be integrated.

ABSOLUTES AND DESIRABLES

The plan implementation leader must bring everything together. He or she must create a type of interrelated system that moves forward in a controlled, predictable fashion. To do this, an eye must always remain on the endgame as current performance is evaluated.

Communicating the plan intent and endgame helps the leader maintain organizational focus. Also, as situations arise and adjustments are made, the clarity and understanding of roles and responsibilities can be updated. The leader is the focal point for identifying accountability for actions, behaviors, and results.

Absolute Must Haves
- *Authority.* The authority of the leader must be formal and accepted by the organization.

- *Clear Roles and Responsibilities.* Clear knowledge of what each team member is expected and tasked to do is a must.

- *Agreement on Expected Outcomes of the Plan.* Agreement and clarity on the desired results, timing, and effects are required.

- *Excellent Sponsorship.* Visible and active sponsorship helps align the organization.

- *Cross-Functional Team.* Balance is a must, as all areas must participate and have representation.

- *Good Data.* Without good data, you are managing air. Everyone has to play by the same rules so the company goals move ahead of any personal agendas.

- *Established Boundaries.* Project scope boundaries must be defined. This facilitates management of activities to keep them in scope.

- *Willingness to Look at Things in New Ways.* As the organization moves forward, new operating and organizational linkages will be required. New standards may be required.

- *Communication.* Establish links both up and down. Build feedback mechanics and paths.

- *Agreed-upon Consequences and Discipline for Follow-Through.* The organization must have the ability to execute and deliver what it promises. It must react to both meeting and missing commitments. Both behavior and outcome expectations must be clear.

What Works Well
- Utilizing cross-functional and multidisciplinary teams
- Utilizing a skilled facilitator
- Utilizing those with program management, lateral leadership and networking skills
- Leading from the front, establishing the accountability of the leaders
- Utilizing active communication

Desirable to Have
- Aligned team and project plans
- Information available to all and timely evaluation
- A burning platform to create focus and motivation
- Communication

Employees must understand the strategies being taken to achieve objectives. Be candid in sharing the realities that are facing the operation and the changes they will cause. Show employees respect by sharing facts and realities. This builds trust, which enables people to engage more freely. Help them understand the effects changes will potentially have on them directly.

This leads into a discussion on the importance of managing change to achieve the endgame.

MANAGING CHANGE

Plans and projects are launched to transform a defined set of activities, functions, or entities toward achieving a specific desired result. Managing change during this process is essential. *It is the single most important component of achieving success in implementing a plan.* Both starting change and keeping it going forward require management.

THE CRITICALITY

Change management is usually the most underestimated need and neglected area of any plan. Compared to managing financial, operational, legal, or sales and marketing activities, the management of change should receive more attention and effort. It usually does not because it is difficult. Results are harder to quantify, communicate, and measure. Effective management of change requires attention, time, patience, and personal investment. Most results show up only after a long period of time. The larger or more complex the organization, the harder change management can be. These groups may have intrarelationships as well as hard established operating practices that undervalue the need for change management.

Change should not be left to evolve on its own. Seasoned leaders report that change management in organizational transformation is the single greatest point needing attention. Without it your business plan will fail! Like managing quality, managing change can be thought of as a fundamental enabler to success. You start with a well-developed quality plan, but your objective is to engage everyone in implementing improvements and integrate the concepts, supportive behaviors, and skills into the fabric of how the business is conducted. The same holds true with management of change. Planning the management of change, in the up-front stage of business planning, will have significant returns along the entire path of execution.

Change is both a condition and a process, according to Sullivan and Harper. As a *condition,* change happens in the overall environment and takes place externally. It is part of the reality outside one's direct control and must be accepted. As a *process,* change consists of the leadership and management actions taken to transform a condition or organization toward a desired objective and takes place internally. It is within one's direct control.

Change is a subset of *transformation.* Most business plans strive to transform the organization or business. Large, aggressive goals and objectives require transformation, not just change. Transformation is best achieved through a structured set of interlinked changes. This allows for achievement in manageable steps, with feedback along the way if adjustments are required.

THE FUNDAMENTALS

Guiding change requires attention to detail and the creation of influence across most or all of the operating horizon. There are two realities to help with this requirement. First, small changes build to big effects. Second, if planning is done well, it can enlist every employee to help in managing change. Both of these help the CEO or key leader in their change management efforts.

UNDERSTANDING SOME BASICS

Malcolm Gladwell provides a unique view of change in *The Tipping Point: How Little Things Can Make a Big Difference* (2000). Tipping point theory suggests some core components of how change is created in the environment that can be used in managing the change process. The theory shows that many small changes add up to substantive change. Change builds on itself.

Gladwell's representation of the tipping theory suggests that there are three characteristics of change:

1. It is contagious.
2. Little causes can have big effects.
3. The point of change happens not gradually but at one dramatic moment.

Gladwell suggests thinking of change as growing and spreading like a disease. It moves from person to person or activity to activity. Ideas and messages can spread by point of contact or also be carried through ancillary vehicles, such as media products. Change happens after repeatedly adding to one side of the fulcrum to tip the scales. A formal plan targeting where to initiate change allows the CEO or key leader to be most effective with their effort and time.

Tipping point theory also puts forth that there are three agents of change: the law of the few, the stickiness factor, and the power of context.

Law of the Few. A few networked, energetic, and enthusiastic people can launch the key messages about change. These are often the thought leaders in an organization and may or may not be the formal leaders. These few people will begin the process of implanting the essence of the change through their activities, influence, and connections. Those who work across horizontal boundaries are important. This law is largely a function of the messenger.

Stickiness Factor. *Stickiness* means that a message makes an impact. Unless people remember what they are told, why would they ever change in the direction desired? People must be able to relate to the message or information. The information age has created a problem, as much of what we read or hear is not remembered. Therefore, it is important to repeat a message many times and through different media. Attention must be given to the various preferences of how people absorb information. Stickiness is largely a function of the message.

Power of Context. Messages are sensitive to the conditions and circumstances of the times and places in which they occur. Tinkering with the smallest details of the immediate surroundings can tip the direction of a change. All actions and messages should be presented in the context of the organization's strategies and objectives. Context is largely a function of the environment.

Use these concepts to increase the effectiveness and efficiency of your change management plans and execution.

OPEN PEOPLE'S MINDS FOR CHANGE

Most people think of change as frightening, and organizations usually act with initial resistance. I have found that it is important to invest time and effort to help people prepare for change before launching any transformational objective. This can be addressed in a number of ways, such as by governance, policy, structural change, factual announcement, or process alterations. I view these as a push strategy for adaptation and do not consider them the total answer. Note that I earlier said, "Invest time and effort to *help people prepare* for change," not *help prepare people* for change. The distinction is important. Change only works if people accept and internalize it. If it is pushed on them, there may always be elements of resistance and lack of ownership.

Managing change by conducting up-front activities that involve key leadership and large sections of the organization leads to great success. Regardless of the activity or exercise, the payoff is significant, and benefits continue far into the future. Activities with the following objectives are recommended.

Objectives
- Promote interdisciplinary participation—more balance creates greater break through solutions.
- Change the operating venue to level the playing field.
- Change the anticipated or practiced organizational rules to open thinking.
- Create situations where existing standards, talents, or understandings may not hold.
- Create a safe environment for learning.
- Use fun, physical effort, or both.
- Force sharing and interdependence to be successful.
- Force new ways of working together.
- Bridge lessons learned and new behaviors directly to business objectives and expectations—this is the most critical component.
- Practice on business situations.
- Debrief to drive lessons learned.

By conducting activities with focus leadership groups before implementing transformation changes, the following benefits are gained.

Benefits
- New skills are developed to address any fear of change.
- Beneficial hidden skills are usually surfaced.
- Engagement is increased due to increased understanding of the *what, why, how, where,* and *when* behind the need for change.
- Confidence in ability to lead change is increased.
- Ability to carry the complete message forward is increased.

This creates a team of lieutenants to help with the change management integration efforts.

Examples.   Comedy improvisation exercises are a great help to prepare for change. Managers and leaders consistently give feedback that these exercises provided some of the most powerful and positive work skills they ever gained. According to Bernie Roehl of ImprovComedy.org, learning the value of gaining agreement is a most critical skill. The performers must agree about location, characters, and events and must learn to accept each other's ideas and build on them.

Learn the value of agreement.   A leadership team is more productive if the members are unified in their direction. This facilitates getting to the endgame goal as a group, pulling in the same direction.

Agree on what is real.   Accepting the conditions and realities that are causing the need for change accelerates activities. Energy and time are not spent on needless debate, and trust is more quickly developed.

Accept and build upon.   Accepting the need for change focuses energy on moving forward and supporting each other toward the endgame. Less time is spent on potential backtracking, needless debate, or personal positioning. Needed corrections can be introduced as the process continues.

Exercises

Yes, And.   Accept and support other players. One must build on what is given and carry it forward. Responses of "Yes, but" or "No, but" are not allowed. Only "Yes [acceptance], and [build, add value, and move forward]" is allowed.

The fundamental rule of this activity is that you accept whatever is given to you, and you cannot change it. It becomes your new reality. This simulates external environmental conditions that one cannot control. The only way to succeed is to build on that situation and move forward. This exercise gives practice at accepting and responding to change.

Heighten and Expand. Take an idea and see where it leads; explore its natural consequences while simultaneously raising the stakes. Then increase the idea both horizontally and vertically.

The structure of this exercise causes the group to explore vertically, as in areas of specialization, and horizontally, as in like markets or areas with common attributes. This simulates exploring the full potential of the business or project impact. The greatest success comes from exploring new ideas together.

So, I Will. Focus on listening skills and get players to take smaller, more logical steps in their story building. Participants must listen, then explain what they understood by paraphrasing and identifying an action with "What you are saying is that...so, I will...."

This is the basic approach to interlinking actions and making them relevant to each person. Major transformation projects such as Six Sigma often seem overwhelming to some. This exercise helps participants translate objectives into digestible steps that each group can relate to and control. It also reinforces the interrelatedness of actions in a balanced plan.

This is offered as an example of how a simple set of activities can be structured to open people's minds to transformational change, developing behaviors that support the changes needed and demonstrating personal investment in the actions to facilitate change.

TOOLS AND ELEMENTS TO SUPPORT MANAGEMENT OF CHANGE

Who wants this change and this plan? Until change directly affects individuals, they will not engage with the need to change or the actions resulting from a change. This holds true for all levels: CEO to employee to customer. It often happens that groups go through the planning actions but do not follow through to change behaviors. This results in suboptimization, complete failure of the project, or lack of sustainability of improvements. Managing change is hard and must target realities as well as perceptions, behaviors, and skills.

Setting the Stage. It is important to raise people's awareness of an oncoming environment of change. This can be done through exercises, as described earlier, or through more structured methods. In all cases you are targeting transformational change through ability, behavior, and action alignment.

Training and Retraining. Determine what skills and behaviors are needed to support the transformation and impart them through formal training. Teach to expose the information, train to exercise use of the information, and educate to develop application of the information. Shape and define behavior through both education and implementation of specific processes.

Apply knowledge management practices to build disciplined processes of activities. The processes will accelerate sharing of information across boundaries. They will also create additive value, as knowledge serves as building blocks.

Pay special attention to program and project sponsors, leaders, and middle managers. Reinforce key skills and behaviors that will allow them to lead the organization confidently. They must be able to answer detailed questions, make decisions on adjustments, and have an appreciation of cross-activity effects.

Establishing Governance. Establish specific assignment of roles and responsibilities. This is best with multidisciplinary participation. Who makes decisions, who is accountable, who is consulted, and who accomplishes the action? Each role is significantly different.

Clearly define the boundaries of the projects. Contract books are used in some organizations to document the boundaries, expectations, and governance of major projects. Each member of the team is required to sign the contract book, thereby committing to the goals, activities, and deliverables of the plan.

Decide whether practice rules or specific tools and methodologies will be used—for example, project management disciplines or tracking and reporting with standardized software. Establishing this type of governance in the early phases accelerates activities later in the process.

Maintaining Alignment.   Work gets done through a network of resources and actions. Getting these resources and actions all pointed in the same direction, sequenced in the correct order, phased for the correct timing and moving at a balanced pace is only part of the challenge. Alignment also requires management of distractions, disruptions, and unanticipated external change. Making strategic adjustments and keeping everything moving forward as planned maintains alignment. The following actions can be used to facilitate alignment.

Conduct stakeholder mapping.   Determine whether roles and objectives are commonly understood. Identify who are supporters and who are detractors. Determine where critical resources or skills may reside. Enlist the help of supporters to help manage the detractors.

Keep the plan current.   Goals and actions must be adjusted to track the endgame. They may need minor change due to feedback, actual performance, or new inputs to the plan. Integrating the goals and performance reporting into the normal fabric of the organization will help to keep it current.

Allocate resources.   Organizations need resources to act. Therefore, limiting or awarding resources only toward specific activities will help create alignment. This *justification* process works if the organization is disciplined in how it allocates resources. Both budget and personnel allocations should go toward those projects that address critical success factors. There should be no hesitation in removing resources from projects that do not. If there is significant resistance, it may indicate that a critical success factor was missed and should be reevaluated.

MANAGING SCOPE CREEP

Definition, measurement, review, and leadership participation are the tools for managing scope creep. Measurement and regular review will highlight issues *if* they are consistently held up against the clearly defined scope boundaries. This takes discipline and documentation. Some organizations use a "project shelf" or "parking lot" for capturing additional opportunities or issues as they are identified. These issues can then be planned for correctly and prioritized for need and benefit against all other projects.

Organizational structures can also be used to help manage scope creep. Maintaining a right-sized resource base for the approved work discourages the expansion of activities. Although lean staffing is not popular, it does instigate collaboration, networking, creativity, and prioritization. Be careful not to become too resource lean and jeopardize the achievement of the desired results of the plan.

Some projects may need to be stopped or postponed. They may be out of activity sequence, imbalanced between investments and return, or an original goal has been adjusted. Shut down any distractive activities!

ALIGNMENT AND INTERLINKING

Ensure that various business units, departments, or key functions are involved in required activities. A matrix identifying key business objectives or initiatives and affected entities can be constructed to ensure alignment and total participation. This can be used to highlight omissions and check for complete cascading of goals (Fig. 2.7).

POSITIVE EXPRESSION

Be consistent in how goal targets are described. Each goal and resulting measurement can be described from both sides of the equation. Beneficial, or value-added, attributes should go up. Negative, or non-value-added, attributes should go down. For example, aim to improve customer satisfaction and aim to decrease defects per unit. It can be subtle but have a large effect on the culture and change the success of the organization.

PROCESS NAME/ DESCRIPTION	PROCESS DELIVERIES, UNITS	TYPICAL OPPORTUNITIES FOR ERROR	MEASUREMENTS Q T C

FIGURE 2.7.   Establishing measurements.

Timely and Routine Reviews

Having to report at a high-level review *always* drives action! The review puts real-time accountability on the presenter. Routine and cyclical reviews motivate actions to keep up the desired pace and keep change moving forward. They serve as a clearinghouse for timely information on performance *as long as* the reports are open, honest, fact and data based, and aligned with the correct goals. They reinforce the lexicon of the plan and the expectations of the leadership team.

Reward and Recognition

This can be a key tool in managing change if used correctly. Reward achievements that really matter in achieving the strategic objectives. Create high visibility in the earliest stages of the plan. This will build momentum for change. Recognize desired behaviors as well as outcomes. Plan sponsorship should be active, visible, and vocal in granting recognition. Highly visible public forums are the best venue for the acknowledgement of significant achievements.

Rewards and recognition should not be given unless a goal is achieved and the results are sustained. Too often, organizations reward individuals or teams for attaining a level only to find that the anticipated savings or benefits are never realized. Establish and publicize a time delay between project completion and award presentation. Therefore, rewards must be based on metric-based measurement.

Collaboration

- Strive to have the outcome of the whole be greater than the possible outcome of the parts.
- Develop processes and champions that act as lateral connectors, identifying and linking positive outcomes.
- Value and bring forward your knowledge brokers, those people throughout the organization who have first-hand knowledge of customers or key processes.

- Cultivate and reinforce the development of a common mind-set and shared understanding.
- Create shared work space and landscapes where real-time exchanges can occur.
- Structure shared goals with dual accountability; let the team establish the approach for action.
- Communicate clear roles and responsibilities, but do not make them restrictive.
- Create *cultural storytelling* sessions; this creates a sense of belonging, transferring values, behaviors, capabilities, expectations, history, and skills among people and groups.
- Create forums for dialogue between levels, such as breakfast with the CEO or brown-bag sessions where open discussion can occur.

LEVERAGING CONTACT POINTS WITH STAKEHOLDERS

Develop linkages between the CEO level and all other levels of the organization. Build forums for both upward and downward communication. Keep as informal as practical and actively listen. Provide the voice of the customer to all employees. This can be done face to face, through media and technology, or through written information. Making the customer a person and the needs real encourages employee engagement.

COMMUNICATION

Communication is probably the greatest basic requirement for successfully moving from vision to results. Providing facts and reasoning to demystify management actions will help people understand the need for change. Communication prowess builds confidence, and confidence leads to trust in the direction of change.

Make Information Clear. Make all communication clear, consistent, unambiguous, and complete so it can be received correctly. Give it life!

Use Repetition. Make the message simple and repeat it. Learning authorities indicate that a message or information must be repeated at least 10 times to ensure that it is understood and retained. Repeat the message constantly and it will be accepted and acted on.

Balance the Big Picture and the Details. Show and communicate information about the endgame; keep it in view. Never let it become totally overshadowed by near-term details or interim messages. Also show the needed level of detail. This can be compared to the way a flying bird surveys the landscape to gain an overall perspective, then swoops in to gain a closer look at a detail on an objective, but may then fly high again to look forward. Communicate information on both your immediate status and your progress toward the endgame.

Communicate What Is Next, and the Next After That. Keep people focused two steps ahead. This communicates direction and intent. If execution of the plan accelerates, the momentum will be carried forward without having to wait for the next instruction.

An example of this comes from the history of the U.S. Civil War. At Antietam Creek, General Ambrose Burnside instructed his troops to take a strategic bridge. The bridge over the narrow creek was the entrance to a strategic city just beyond the immediate hill. After a long and difficult battle, Burnside's troops took the bridge and stopped. They did as instructed and took the bridge only. The next step or intent had not been communicated. This delay allowed the enemy to regroup, retrench, inflict heavy losses, and significantly delay Burnside's next forward motion.

Respect Different Learning Styles and Perspectives. People learn and absorb information in different ways. The diversity of a workplace has people coming from different points of understanding. There are visual, verbal, and written communication preferences. Your communications should cover them all. Present information by various means, thereby canvassing the entire audience. Increase the opportunities for people to understand and gain knowledge.

There may also be cultural perception differences. Translation of a manufacturing document for a global operation resulted in the verb *ship* being translated into 12 disassociated words. People could not act when told to "boat" a product. Search out feedback to make sure that the message has been understood.

Be Factual. Sharing information and management facts (such as shipments and forecasts) prevents misinformation, rumors, and counterproductive perceptions from distracting employee efforts.

Be Persuasive. Communication has an element of marketing. Answer known and anticipated questions in your communications. Build a case for action. Inspire and be positive.

Be Pervasive. Integrate your message in all communication and media opportunities. This includes formal and informal opportunities. Include both internal employees and external stakeholders.

Use Feedback Mechanisms—Listen. Establish feedback loops both up and down. Identify and manage barriers to communication. Value input from all levels.

The U.S. Army developed a powerful feedback process called an *after-action review.* Immediately after a simulated battle, key individuals from *every* level of activity come together in a structured process to evaluate what went right, what went wrong, and what could be changed to make improvements toward the end goal. In this forum, the lowliest private has the same valued voice as the highest general. Feedback may include the general saying "You missed..." and the private saying "You did not know that...." The inputs are treated positively and intended only to improve the process, identify areas for improvement, and increase the probability of success.

Areas to Consider for Communication
• Written—active versus passive styles
• Oral—speaking and listening
• Visual—such as learning maps

- Internally and externally focused
- Formal (reports) and informal (ad hoc conversations)
- Media based
- Presentation techniques—visuals, body language
- Meeting management techniques—integrate into agendas
- Vertical and horizontal directions

Make change relevant and real. Talk about it, act on it, and integrate it into your business activities. The goal of managing change is to have each and every employee, regardless of level or function, feel that they are being affected by the initiative or plan objectives. They must understand the *why* behind the need for change and be compelled to take action. Each person must understand how their role and actions relate to the company goals and desired results. They must also know the current status, know what steps to take next, and have a clear understanding of the endgame.

ACHIEVING RESULTS

If you *say* it, *do* it! This is where actions must bridge the span between the plan and the desired results. Clear goals, prioritized actions, measurement, auditing, and refinements get the organization across the gap. Failure to achieve results is usually due to poor organizational execution processes rather than the design of the plan and the structure of the objectives.

The effects of failure have two dimensions. First, the organization has invested significant time (6 to 9 months usually) in resources, time, and attention without the desired outcome. Effort and investment have resulted in being in the same place without return on the investment. Second, the competitive landscape has shifted. Competitors, technology, and other disruptors have moved forward *at least* 6 to 9 months. Your investment has been made, and you have gained no return, while the competition has leaped ahead of you by a significant distance.

You must manage to achieve the desired results and not allow failure.

It is usually recommended that the sequence of activities be launched to achieve some small, quick wins. This starts momentum and builds excitement and buy-in to activities. Projects and actions that are fast, easy, and controlled at the team level are good candidates. But each of these must be aligned to the higher-level plan.

Celebrate and market successful performance. It is important to reinforce successful actions, especially in the early stages of a plan deployment. Recognize achievement and those who contributed to that achievement. *Caution:* Pay attention to recognize *all* contributors to increase alignment, common mindset, and unified activities.

GOALS, COMMITMENTS, AND RESULTS

These three items should make a tightly closed loop. They must be sequentially linked for success.

Goals. Clearly understood organizational and process goals will help create a common mindset and common language across the company. The use of a *SMART* goal structure provides the components of a well-structured, easy-to-communicate, and measurable goal. Communication of goals must be consistent and frequent. When employees understand the goal, it empowers them to take action and make personal investment of their talents.

Commitments. An organization must demonstrate the discipline to track commitments and the accountability to act on what it commits to do. This creates credibility and reinforces expectations. Formal tracking can be accomplished using any number of management tools. One example is a simple multi-columned chart, as shown in Fig. 2.8.

The real value in achieving results comes from the process behind this chart and the organizational alignment it creates. Tracking and reporting can be integrated into normal operations or business meetings. Keep formats simple and fact based. Communicate the consequences of achieving or missing commitments that are made.

Results.   Measuring and reporting results makes your progress toward attaining your goals visible. Guidelines for measurements include the following.

• Keep measurements simple and have only a few goals. It is recommended that you have no more than 2 to 4 goals for each area of focus.

• Track key process indicators. These are the early warning signs of issues, allowing time to initiate corrective actions or refine activities. They indicate whether you are on the path to achieving the desired results.

• Streamline assimilation of information and evaluation of status by using structured metrics and dashboards. Focus on the few key measurements for each focused area.

• Use measurement, data, and facts to create a common language about activities and document expectations.

Make measurement and reporting a shared responsibility. This will increase the involvement and investment of employees. However, it is best if an organization can dedicate resources for the *management* of data critical to that measurement and reporting. This provides a neutral party, objective facilitator, and reporter of results. This approach knocks down barriers and increases alignment between groups.

OBJECTIVES	DEPARTMENT 1	DEPARTMENT 2	DEPARTMENT 3	DEPARTMENT 4
Objective 1	Goal	N/A	Goal	Goal
Objective 2	Goal	Goal	N/A	Goal
Objective 3	Goal	Needed	Goal	Goal

FIGURE 2.8.   Alignment matrix.

Say				Do	
GOAL	PERSON COMMITTING	COMMITMENT	COMMITTED DATE	COMPLETED DATE	RATE OF ACHIEVEMENT

FIGURE 2.9. Commitment chart.

AUDIT—A MUST!

Check results. Test along the way. Evaluate effects between goal activities. Make required adjustments and provide feedback. Audits of activities can be integrated into normal business processes such as SEI evaluations. The more the high-level plan becomes a driver for normal processes, the faster the plan activities become reality.

COMMUNICATION

The objective of communication, in achieving results, is to provide:

The right level of data

At the right time

In the right format

To the person or team

Who must take action(s)

Required to meet the goals.

SUSTAINING AND ENHANCING PERFORMANCE

The job is not over once initial results are achieved. Institutionalization and transition of improvements into the ongoing performance is where the real power lies. Results should be sustainable to have an effect and actually be realized. Every improvement raises the bar of expectations, so

improvement must be sustained to remain in the same competitive position. Management and leadership activities must
continue. Aim for both change and rate of change in achieving results. Build the organizational muscle to stretch further
each time. This is a journey, not a day trip. Transformations do
not happen quickly, and they stick!

FOLLOW THROUGH ON ACCOUNTABILITY

The first step in moving forward is to do what you say.
Reinforce expectations on outcomes and behaviors.
Acknowledge both hits and misses.

COMMUNICATE RESULTS

Measure performance and formally communicate results on a
regular and timely basis. Be positive but objective. Look for
achievements, not failures. Use data, metrics, and simple
reports. Issue summary reports at key milestone points to communicate how far along the path toward success the company
has come.

KNOW WHEN TO MAKE ADJUSTMENTS

As the plan unfolds and the entire team engages, changes will
need to be made. These changes are not snap decisions made
in the moment or mere reactions to current events or performance. Changes should go back to the strategic plans and
objectives. Trigger points on key indicators should have been
identified through scenario analysis. Flexibility was designed
into the plan, along with situation response options. Be disciplined in following these adjustment triggers to keep moving
toward the endgame. Maintain flexibility to react to changes in
the landscape.

MAINTAIN YOUR TOOLS

Just like a piece of capital equipment, you need to maintain the
investment you have created through your plan with periodic

updates and review. Refresh and renew systems for advances that may occur during the execution of your plan. Build on the core processes and make them stronger. Use them to remain flexible and focused.

The following lists highlight some of the tools required to execute your plan. Keep them up to date and running well.

Measurements (Simple and Few)
    Dashboards and metrics

    Key process indicators

    Key performance measures

    Auditing

Communication
    Lexicon

    Feedback processes

    Knowledge management processes and sharing

    Communication of status

    Reinforcement of mission, vision, strategies, intent, and goals

    Reward and recognition standards and events

    Celebration of attaining milestones

Policies and Governance
    Update of roles and responsibilities

    Structure of projects

    Integration into emerging projects and activities

Leadership
    Visibility, vocal communication

    Leadership from the front

Use Stretch Performance Motivators—Power in a Simple Symbol. Here are three real-life examples that demonstrate the fundamentals of sustaining and enhancing performance. There is wonderful power in using even the simplest item as a symbol if it is given context within your plan.

Example 1.

GOLDEN BANANA — MOTIVATING FOR REACH-OUT GOALS

A friend, Al Filardo, related this story.

> A manager was so overjoyed with the performance of an employee that he wanted to reward him immediately and recognize the achievement on the spot. There was little in the meeting room other than a banana from the refreshment tray. The manager picked up the ripe banana and announced, "I award you this *golden banana* of achievement." Although it created a laugh, the manager's genuine recognition and respect for the accomplishment, from a very personal level, showed through to the employee. He kept the banana until it went bad and regretted having to throw it away. *It wasn't the banana that mattered—it was what the banana represented, and that lived on.* Both the manager and the employee related the Golden Banana award story when discussing the achievement. Eventually, the banana came to symbolize exceptional performance within the company, and employees referred to "banana" situations within their business activities. The symbol became so powerful that artificial golden bananas were eventually produced, and the manager awarded them only for ever-increasing levels of accomplishment.

This example demonstrates:

Using recognition and reward
Communicating and creating a lexicon
Establishing the expectation of reach-out goals
Instilling fun

A small thing, spontaneous action, and context provided by the manager combined into a motivating image. Simple symbols can also be used to cultivate collaboration in highly complex operations. They can become a motivator for achieving the highest of customer and business objectives in extremely large groups.

Example 2.

INFORMATION AND
A DONUT — COLLABORATION
TO ATTAIN ALL GOALS

In this example, a *donut* became the symbol of achievement for an entire global product and manufacturing organization (over 3300 employees). A very forward acting general manager named Rick Chandler created this system in the early 1990s. I had the privilege of carrying it forward and expanding its power.

> The business was producing four different generations of highly complex telecommunication base stations and switching systems. It was during a highly competitive period with rapidly expanding global market requirements. Therefore, delivery of systems was the highest customer requirement and became our prioritized focus.

> Only achievement of *all* goals determined success. We balanced the equipment delivery focus with reporting product and service quality, operating and cost efficiency, new product introduction planning, and employee development goals and performance. It was amazing how quickly all levels of employees came to understand the interrelationships between quality, cycle time, cost and delivery performance, and the resulting customer satisfaction and new order activity. They also developed an appreciation for the need for investment in future activities like innovation and systems support.

> Sharing business and performance information in a simple structured format increased understanding and empowered each employee to make better decisions in their daily routine. The structure was repeated at each facility around the global operation, making minor *additions* to relate more closely to the respective organization. Nothing was omitted from the core format information. Everything linked to the ultimate highest divisional level goals.

> The structure was simple: If all groups met their goals *and* the total organization met its goal, everyone enjoyed coffee *and* donuts during the business meeting. If a single group missed their internal goal, *even if* the organization achieved the higher-level goal, only coffee was served at the meetings. This facilitated a culture of collaboration. A group could indicate it needed help or had hit a barrier, and the entire

organization would bring resources to address the issue. It became amazing, as senior managers would use "Donuts may be in jeopardy this month!" as a rallying cry. Everyone pulled together for the good of the greater whole. This included achievement of the interim goals (critical for next month's success).

Counter to the culture at the time, our division created a monthly business communication meeting to share critical information with *all manufacturing employees*. It became known as the donut meeting. The intent was to:

- Focus attention on commitments versus actual performance.
- Communicate and reaffirm the big picture.
- Create an understanding of the need for balanced performance.
- Establish organizational expectations.
- Confirm both short (monthly) and interim (quarterly) customer performance goals.
- Create relevancy so each person knew how their role impacted the ultimate goals.
- Reinforce a common mindset.
- Encourage collaborative support between groups.
- Provide customer feedback.
- Provide a platform for current issues, information, and policy communication.
- Conduct recognition and reward.
- Have fun.

All of this was accomplished in a 30-minute meeting conducted monthly in the company general cafeteria with very low-tech presentations. The quarterly review was extended to 45 minutes. Investment for the "reward" donuts? Less than 35 cents per employee: very cost effective!

The general manager, or head of each facility, was provided a platform to demonstrate leadership and personal investment in the operations and its people. We became the face and voice of the business and strategic plan. Attendance increased dramatically as other groups, sister divisions, and even vendors began to attend. When asked why they came to a manufacturing meeting, they indicated it was the only place for them to gain insight on the activities and plans of the entire operations. It was the balance of reporting and disclosure of near-term goals that benefited them in doing their own jobs that caused them to come. Questions were originally raised about openly sharing forecasts with vendors and other external groups. I took the position that vendors needed the visibility to partner and support our efforts; we made them part of the equation. It benefited the operation many times over.

Minor adjustments to the format and food (donuts don't work in some Asian facilities) were made to reflect cultural and location differences. The core intent, integrity, and content of the meeting were always maintained. They provided open communication about goals, commitments, and performance. This meeting format became a business best-practice and was implemented in other sectors and divisions.

Example 3.

TRAINING, LEARNING, AND INVOLVEMENT — TRANSFORMATION OF AN ENTRENCHED ORGANIZATION

Transformational improvements are not limited to areas of heavy investment. Even an end-of-life product with extremely limited resources can transform itself and its people.

> The transition from analog to digital technology was complex, expensive, and compressed. A sister division was in trouble and needed funds to invest toward meeting customer and new digital business requirements. Our division manufactured the cash cow product, but it was at the very end of its life cycle. Quality had started to slip, and project funds were basically nonexistent. Funds were directed toward the strategic growth of the newer product. Yet there was a common customer base, and many of these customers had been targeted for the new product introduction. We had immediate quality and dollar objectives.

There was a need to motivate an extremely large manufacturing and accompanying small support engineering organization to improve performance, in a short period, on a dying product without any major monetary investment.

Challenging? Yes! Here is what we did.

Analysis.   The structure of each product was broken down into its sellable components. Each process was matrixed against each component. Teams of manufacturing employees were established to evaluate opportunities for improvement at each intersection point of the matrix. Every process and every part had coverage. Everyone participated! Experts began to volunteer their time where needed on special issues.

Enabling the Analysis.   A list of needed skills and understandings had been developed before the teams were drafted. Training began on the fundamentals of profit-and-loss (P&L) structures, how to write a simple business case, ABC cost analysis management, and the fundamentals of Six Sigma. This covered cost, quality, cycle time, and skill development. Each tool was set up to mandate analysis of each attribute.

Project Activities.   Managers were assigned mentorship roles with each team. Potential projects were identified and analyzed using the standardized tools that had been provided. Then report-out sessions were held. Note that work did not start! We broke activities into two steps: (1) generating and evaluating opportunities, and (2) acting on opportunities.

Report-Out.   The training and skills that had been developed were significant, and we wanted to recognize that achievement. The level of integrated involvement was record-breaking. Ideas were being generated everywhere and on everything but were controlled through the standardized tools we had developed. We wanted to recognize the value of this effort.

We also wanted to reinforce the importance of prioritizing where we expended our limited resources in order to get the biggest return. Therefore, we established a type of proposal review meeting. Teams submitted *all* the business case reports on ideas for improvement. A leadership team member reviewed each proposal. Because there were so many, six were randomly picked for presentation at a monthly review meeting.

The Magic.   To reinforce employee engagement in the business, we turned these sessions into recognition sessions. We celebrated the analysis of opportunity and established an approval gate to move projects forward. Proposals that created improvements were celebrated; proposals that determined what originally looked like good ideas but were actually not beneficial were equally valued and celebrated. The reports were limited to 5 minutes each, so a lot of information was shared in a short period. We then included reports from teams where project work was proceeding. This gave real-life extension to the previous training.

Probably the most important thing we did was to create visibility and feedback with the sector president. We were able to secure his attendance at the report-out sessions. Manufacturing employees were overjoyed to have an intimate audience with him to present their work. The president was overjoyed to see the level of productive work and impressive end results. He fed off the energy of the group.

Balanced Results.   In less than a year, 3,300 employees had learned the fundamentals of P&L structures, Six Sigma, ABC activity management, and business case analysis. They had also gained communication, teamwork, presentation, and platform skills. The end-of-life product simultaneously gained improvements of 30% in quality, 20% in cost, and 22% in cycle time with only minimal investments of funds. Cash flow was greatly increased and provided resources to the new digital product development group.

In addition, so many improvement ideas were generated and evaluated that we actually developed a project shelf. The highest priority was given to projects delivering the greatest return across the balance of all goals. We were able to sequence and align the projects with activities in sister divisions. The knowledge gained from several projects was advanced into the new product designs. People also advanced into new, more valued roles based on the skills they had gained and further investment in their capabilities.

This example demonstrates:

Recognition and reward.   Reinforce the behavior that leads most directly to the goal.

Communication and creation of a lexicon.   Everything was data driven and used standardized formats and analysis processes.

Engagement of employees.   People began to bid on joining new teams.

Instilling fun.   Review meetings had a theme that kept things light.

Role of leadership.   Attention was highly motivating; information was rewarding to the president.

Knowledge sharing.   Opportunities were evaluated and/or solved once; records were kept; ideas were exchanged with a common lexicon and in structured ways.

Sustaining and enhancing performance is highly rewarding to a leader. It requires continued attention and support, making necessary adjustments and keeping the organization focused on the endgame through prioritized strategies.

AVOID BLIND SPOTS

Even the most complete and best-managed plan may have blind spots. Even the best organizations and actions may still hit significant snags. Experiences from developing and implementing plans, both large and small, point to areas of potential concern. Those who have accomplished repeated cycles of implementing plans provide the valuable comments and strategic and institutional lessons highlighted here. These are frequently missed, with negative consequences.

These items assume that congruent team activities, honest communication between all the levels, and organizational planning activities that are both solid and supportive are in place. Each area can create a dangerous blind side if not addressed. Some require awareness by the CEO or key leader of his or her own actions. Some require that the leadership question and manage specific activities. Consider the effects on achieving your desired results if these items are ignored. They will usually have a very negative effect on your plan.

Several areas of caution are highlighted. These have been generically and consistently reported from industries spanning retail to pharmacology to education to telecommunications to medical services. They cover functions of manufacturing, product development, sales and marketing, information technology, operations, customer services, teaching, and patient care. They are independent of business or operation type. They apply to an even greater degree in unique situations or structures.

RECOGNIZE INFORMAL BOUNDARIES

Formal organizational and operational boundaries are documented and should be acknowledged in your planning process. Equally important, and *most often missed*, are the informal boundaries. These are critical to your plan success. More than politics, informal boundaries dictate how work is really accomplished and where power really resides. They can have the most potential effect, both positively and negatively, on achieving results.

Identifying informal boundaries requires questioning and observation. How is work really conducted? Who influences whom and what? Which processes change during a crisis? Are there pending regulator standards looming? Are there groups whose work never connects or converges and why not? Have you accounted for multinational and global differences? Are there potentials for cultural conflicts? Do you have a confident understanding of the mechanics of power within your company?

There are strong working bonds and informal boundaries identified in the answers. Take them seriously!

Make sure to also account for these relationships and informal boundaries in the objectives, goals, and tactical plans. Leverage the ones that help and manage the ones that might hinder efforts. Help team members understand and identify all the boundaries. Support employees to eliminate, manage, or overcome those boundaries that are detrimental.

DO NOT ASSUME HORIZONTAL CONNECTIONS

Many leaders make the assumption that people will make the required horizontal connections on their own. Your view, by design as the leader, allows you to see connections needed for success. Leverage points, communication requirements, and needed working relationships may seem obvious. Too often the assumption is made that "I understand it and the action to take is clear to me, therefore I assume it is clear to everyone else." This does not always hold true.

You have to test for understanding! Have you assumed that critical objectives moved forward because everyone knows how to do that or knows who to work with? Your abilities, innovation, resourcefulness, or experience differ and may even exceed those of others on your team. You participated in development of the balanced plan and understand the required interrelated links. Other managers and team members have not been exposed to that same level of detail. They may not have your management skill or experience level. They may proceed on what they know, believing they have completed the required actions, and yet miss critical connections required for plan success.

Recognize that part of your leadership role is to mentor, explain, and support. Ask and listen! Don't assume! Make sure that important horizontal connections are being made. Build in actions that test for required connections. Probe for the health of these connections in your formal review process.

PREVENT OVERDELEGATION OF ISSUES

Part of the leader's role is to delegate responsibilities and actions appropriately. Delegation can go too far in some situations. Of course, what and how much is delegated depends on the skill level of your team, experiential growth objectives, capabilities, workloads, and similar issues. In plan development and execution, there are key areas where leaders and CEOs should be cautious about how they delegate responsibilities.

A frequently related issue is when a CEO expounds on a strategic objective, followed by exhortations to "Just make it happen!" or "Figure it out!" Such CEOs effectively remove themselves from the planning process before the critical objective and goal-setting phase. Goals for achieving results *will* be designed by those participating. These *may not* be goals for the *desired* results that support the company plan, in line with the CEO's understanding.

CEOs may also disengage their own attention at a critical stage, imparting their authority to another leader. Without adequate preparation and communication, this can be viewed as the CEO being distracted or losing confidence in the plan. With appropriate methodologies and systems put in place, delegation can work effectively. Effects on activities can be minimized if the new leader is adequately prepared, is recognized to have the same mindset, and continues strong feedback with real-time communication with the CEO.

Delegating responsibility and authority also transfers an important organizational reality: influencing the organization. Influencing the organization is a direct link to creating the ability to get things done. If too much is delegated, CEOs may lose their ability to influence the organization and motivate required actions.

Overdelegation can also begin a chain reaction leading to creation of an overly bureaucratic structure. Additional levels for approval, layers of involvement, or unnecessary cycles of review can slow progress. They also can complicate actions and cause frustration for those acting to achieve goals.

Part of the leadership role is to delegate activities and focus on monitoring actions. Yet, overdelegation can remove the CEO from critical points of the plan activity.

MANAGE REQUIRED SHIFTS IN SKILL MIX SETS

Development and execution of a balanced plan requires a mix of strategic and tactical skills. Rarely are individuals strong in both skill suits. Utilizing the correct skills at each stage of activity requires an operating and management environment that values and leverages each skill type at the appropriate step.

Critical contributions made by an individual at the strategy development phase may build confidence in the team's perception of that person's value. Leaders frequently continue to endorse that member's contribution as activities move to the tactical planning and execution phases. This can exclude or limit contributions from team members with stronger tactical experience and skills. You probably do not want the travel agent who planned your vacation to actually pilot the plane that takes you to the destination!

Creating a working environment where people with each skill type can contribute their strong suits, at the appropriate phase, is a critical responsibility for the leader. Demonstrating open support promotes a culture that values and endorses diverse skills. This in turn cultivates stronger common team behavior.

All members should continue on the team for the duration of the activities. Each skill type is needed throughout plan development and execution. Their time should ebb and flow to match the activity needs. Attention must be given to balance skill sets and leverage each contribution at the right time.

MANAGE FEARS AND OWNERSHIP OF THE CURRENT STATUS

Human nature, especially in the business environment, creates a desire to belong. Each person has a need to identify with an event, a success, an organizational unit, or an operating culture. People will own this identifying factor, often enhancing the image to perfection in their minds. They own *it*, protect *it*, and will defend *it*. This creates a breeding ground to increase the natural fear of change and fear of the unknown. Frequently, development and execution of a balanced plan may be perceived as threatening to the status quo, a past success, a style of management, or the very nature of operations.

By design, the plan is targeted on the creation of transformational change. The very skills, styles, and behaviors that made managers or operations successful in the past may be the opposite of those needed in a balanced plan approach. Heroics, singular contributions, advancing more rapidly, or beating the plan can all cause problems. The very thing that caused a person or organization to advance may have to change for the success of the whole. This can subconsciously be threatening to the leaders, managers, and team members who must develop and execute this new plan. This fear may be exhibited in subtle ways, through unjustified hesitation, continual backtracking, excessive critiquing, or aggressive challenges.

Often, people do not realize that they are defending their status or the status quo. Leadership must create an operating environment focused on the total business, where the need for change is understood, encouraged, and engaging. Focusing on desired results, using data measurement, communicating expectations, and following through on accountability will help people move forward. Creating excitement about the future state will also help people make the transition more quickly.

AVOID FIXING UNIQUE ISSUES

While ownership is a desirable trait, reuse and cross-implementation of solutions are even more desirable and beneficial to the business. Most businesses discover that similar problems have been solved many times by various plans and teams. Usually not intentional, this may be a result of closeness to the situation. The desire to put a spin on a plan, making it specific for a particular objective, can become almost an emotional issue for some management teams. Creating a new plan of action may be considered more expedient than spending time to research reuse of a prior solution or plan. It is usually far more beneficial to research and reuse wherever possible.

A balanced plan should benefit the total business. Therefore, opportunities for larger scale, interdisciplinary solutions should be leveraged. Do not allow reinvention of the solution or plan for each area. Manage the use of activities and solutions to bring the organization together.

ENCOURAGE SHARING BAD NEWS

An operating culture based on measurement of data and results is powerful. Both the good and the bad are brought to the surface for business leaders to address. How leadership reacts to these facts, especially challenges or bad news, can shape what information is presented to them in the future.

Managers and workers want to do their jobs and want to be successful. Through their actions and reactions to what is presented, leaders always establish what information will be shared with them going forward. Where the reaction is negative or hostile, managers may sandwich bad news between good news, hoping to minimize the visibility. If the CEO is generally optimistic by nature, almost a required trait, he or she may not hear the bad news. Acknowledging bad news allows you to react and get the plan back on track.

Relying on data and measurement are again part of the solution. Leadership reaction and actions are the critical components. Establish a review environment that is business and fact focused. Draw out details, listen, and engage participation in the development of solutions.

APPLICATION OF SIX SIGMA TOOLS

SIX SIGMA METHODOLOGY

Many corporations, large and small, have adopted Six Sigma to improve their performance. Many large corporations have published their success stories. However, very few small companies have published success stories reporting tremendous savings. There are several reports of successful projects saving thousands of dollars, but nothing like the savings reported by large corporations. This disparity has caused an illusion that Six Sigma is for large corporations. The reality is that there are millions of small to medium-sized companies that can benefit from the Six Sigma methodology. The Six Sigma methodology must be adapted to the needs of smaller companies.

Six Sigma methodology consists of tools that are learned over many years. The tools range from graphical tools like Pareto charts to advanced statistical tools like response surface methodology (RSM). Large corporations have internal resources with varying degree of expertise, while small companies may not have sufficient internal resources to support the Six Sigma initiative.

ROAD MAP TO SIX SIGMA

Most Six Sigma initiatives to date have been project oriented. Typically, implementing Six Sigma methodology on a project leads to some improvement; however, the company doesn't realize the full benefit of implementing Six Sigma. In order to maximize the effect of Six Sigma, a corporation must commit to the implementation in its entirety. In order to implement the Six Sigma initiative, the company must gather necessary information and have the executive team be trained in the intent of Six Sigma. Six Sigma is not about quality improvement techniques; neither is it about statistics. Instead, Six Sigma must be treated as a strategic initiative to improve profitability and accelerate growth. With such understanding, the chief executive must make the commitment to adopt Six Sigma. The road map to Six Sigma is shown in Fig. 3.1.

One of the major challenges is managing investment in the Six Sigma initiative. The executive must make the investment commensurate with the opportunity for profit and growth. The main difference between Six Sigma and similar improvement initiatives is that Six Sigma must be utilized when the improvement in the bottom line becomes apparent. With a clear understanding of the opportunity, the executive team must establish needs and goals clearly. If the company decides to improve its financials, the goals must be set accordingly. If the objective is to reduce waste by improving processes, the goals must be set to achieve Six Sigma–level performance for various processes.

Corporations try to develop in-house expertise to implement Six Sigma. Initially, outside help can prevent false starts and facilitate development of a corporate vision and strategic plan. Then a leader (Sponsor) and the first Black Belt are identified to lead the effort. The Sponsor and Black Belt develop a business model, identify growth and profit streams, and list opportunities for improvement. These opportunities are prioritized and used to define various projects. The projects are converted into an opportunity for improvement. The cost, savings, probability of success, and time to complete are used to prioritized projects. The prioritizing criteria can vary from company to company, depending upon the business objectives.

Once the projects are identified, qualified candidates for Black Belt or Green Belt training can be identified. Some Black Belts rely heavily on statistical techniques, others more on

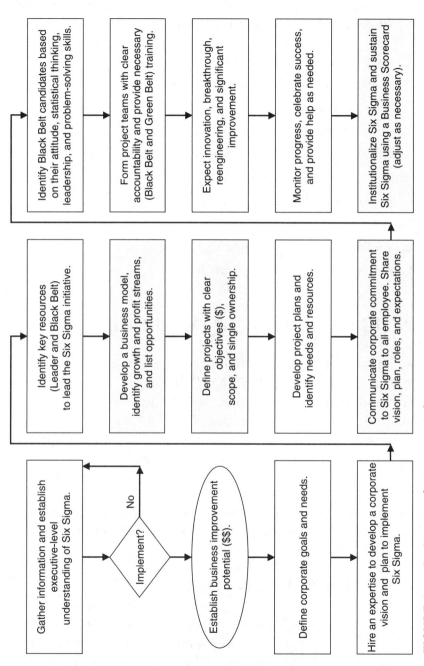

FIGURE 3.1. Six Sigma implementation road map.

93

engineering techniques. For a Black Belt to be successful, one must look into the following Black Belt attributes:

- Positive and winning attitude
- Business and personal common sense
- Statistical thinking
- Leadership and facilitation skills
- Training in tools and techniques
- Aptitude and experience in problem solving
- Innovative problem-solving thinking
- Broad interests and hobbies
- Cross-functional experience
- Curiosity and passion for learning

SIX SIGMA BODY OF KNOWLEDGE

ASQ has established the certified Six Sigma Black Belt examination to qualify candidates for Black Belt. One of the requirements is some minimum project experience. A summary of the Body of Knowledge is shown in Fig. 3.2. One can see that the Body of Knowledge includes areas such as enterprisewide deployment, business process management, and project management in addition to Define, Measure, Analyze, Improve, and Control (DMAIC), Lean, and Design for Six Sigma. The Body of Knowledge list is quite exhaustive, and no one can learn all these tools effectively in the application sense in a short time.

KEY BLACK BELT TOOLS

Based on the Body of Knowledge for Black Belt certification, Fig. 3.3 identifies the key Black Belt tools. The tools are arranged according to the DMAIC phases. The figure includes many tools as well as methods. Not all the tools are used to solve a given problem, nor can one tool or tool set solve all the problems.

Elements	Knowledge Requirements
Enterprisewide deployment	Value, systems and processes, leadership roles and responsibilities, organizational goals and objectives, key metrics and scorecards, project selection process
Business process management	Process vs. functional view, project measures, voice of the customer, customer require ments, parts per million (PPM), defects per million opportunities (DPMO), defects per unit (DPU), rolled throughput yield (RTY), cost of poor quality (COPQ), benchmarking
Project management	Project charter, planning tools, team leadership, team tools, managing change, organiza-tional roadblocks, communication
Define	Project scope, top-level process maps, metrics, problem statement, baseline and improvement goals
Measure	Process analysis and documentation, descriptive and inferential statistics, measurement systems analysis, C_p, C_{pk}, P_p, P_{pk}, C_{pm}, Sigma levels
Analyze	Multivariable studies; regression analysis hypothesis testing; ANOVA; tests for means, variances, and proportions; nonparametric tests
Improve	Design of experiments (full-factorial and Taguchi), Significance of results, response sur-face methodology, evolutionary operations (EVOP)
Control	Statistical process control, rational subgrouping, control charts, precontrol, short-run SPC, EWMA, measurement system reanalysis
Lean and DFSS	Lean concepts, theory of constraints, value chain, Lean tools, QFD, robust design, failure mode and effects analysis (FMEA), DFM, Inventive problem solving (TRIZ)

FIGURE 3.2. Body of Knowledge for ASQ's Black Belt certification.

Experience shows that in order to solve problems, the Black Belt must have the ability to learn a new process quickly, analyze the data using various tools, work with people in the area responsible for the process and seek their ideas, identify all the pieces of the problem puzzle, and then facilitate the solution.

In one example, a Black Belt wanted a team of inspectors to perform a Gage Repeatability and Reproducibility (Gage R&R) assessment for a new product. He prepared a document with detailed instructions for the inspectors to follow, gave the document to the inspectors, and expected them to complete the Gage R&R assessment in the next couple of days. After a while, nothing had happened. Investigating further, he learned that, first of all, inspectors are just as busy as engineers. Second, the inspectors did not understand the document, and no one had explained the Gage R&R method to them. Most important, they felt they were not treated respectfully. Bottom line, the assessment was not completed.

In another situation, when a problem was identified, the engineer asked the operators about their understanding of the problem. They knew the most likely cause and the remedial action. The problem was solved promptly, and the improvement realized was greater than 50 percent. When the Black Belt tools are applied with care for people and with statistical thinking, great results can be achieved. On the other hand, when Black Belt tools are applied without care for people and without statistical thinking, one can expect lots of data and analysis without any solution in sight. In many cases, the Black Belts conclude that the designed experiment is inconclusive, and the solution is not found.

The lesson is that the Black Belt tools are to be utilized with understanding and care. Statistical thinking is more critical to the Black Belt's success than knowledge of all the statistical tools.

DMAIC METHODOLOGY

Prior to implementing the DMAIC methodology, Motorola had a five-phase problem-solving methodology, similar to Ford's 8-

Define	Measure	Analyze	Improve	Control
Surveys, storyboarding, focus groups	Terminology	Plan, Do, Check, Act (PDCA) cycle	Triz	Leadership, facilitation skills
Affinity diagram	Cost of poor quality (COPQ)	Scatter diagram, stem and leaf, box plots	Comparative experiments, components search	Process thinking
Kano analysis/QFD	Normal distribution (random vs. assignable)	Case-and-effect analysis, fault tree analysis	Full factorial, fractional factorial, Taguchi methods and experiments	Precontrol charts
Pareto chart	Measurement system analysis	Multivariable analysis	Response surface methodology for optimization	Control charts (attribute and variable)
Process mapping supplier, input, process, output, and customer (SIPOC)	C_p, C_{pk}, P_p, P_{pk}, DPU, DPMO, rolled throughput yield	Failure modes and effects analysis, tree diagram	Nonparametric tests	Internal audits, corrective action, and management review processes
Critical to quality (CTQ)	Probability distributions	Regression analysis	Advanced statistics	Six Sigma Business Scorecard
Force field analysis	Sampling	Testing of hypothesis	Lean thinking and tools	
Project charter		Analysis of variance	Team-building skills	

FIGURE 3.3. Key Black Belt tools.

D problem-solving approach or even to DMAIC itself. When Motorola initially implemented Six Sigma, DMAIC did not exist in its current form. The three levels of statistical tools training included data collection and display techniques for operators, descriptive statistics and comparative experiments for manufacturing engineers, and advanced statistical techniques such as design of experiments (DOE) and RSM for solving tough problems. The DMAIC methodology incorporates an extensive tool set beyond the statistical tools. It focuses on the voice of the customer, stakeholder analysis, supplier, input, process, output, and consumer (SIPOC), Triz, project management, and Lean tools. The five phases of DMAIC highlight various aspects of problem solving, as shown in Fig. 3.4.

A project is selected for improvement by the DMAIC methodology based on a review of business performance. The Define phase clarifies what problem needs to be solved. The Measure phase enables you to establish the baseline process capability. The baseline may need to be established in terms of the input parameters, in-process parameters, or output parameters. The Analyze phase allows you to identify sources of defects, problems, or variation. The intent of the Improve phase is to achieve significant improvement through employee participation and innovation. Finally, the Control phase allows you to sustain the gain and improve further.

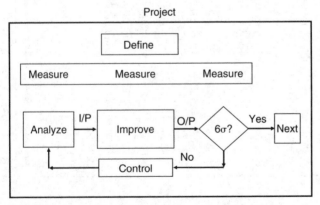

FIGURE 3.4. The Define, Measure, Analyze, Improve, and Control (DMAIC) process.

Figure 3.5 highlights details of the DMAIC process. In the figure, each box has been assigned to a phase of the DMAIC methodology. Notice that the Define phase is the most significant step. If the Define phase is not very effective, the solution may be inconsistent with respect to the project objectives.

The DMAIC road map starts with writing a clear problem summary or project statement. The customer's input is considered in understanding the effects and scope of the problem. The team is formed, and the members identify stakeholders, their support, and their goals. Team members make the business case for a better understanding of the economic effects and for defining the required resources. At the end of the Define phase, the team has a clear project charter. The team constructs a process map for the scope of the problem, identifying areas of inconsistency, and constructs a SIPOC analysis to identify players in the theater of operations.

During the Measure phase, the team establishes measures of variability that describe the symptoms of the problem. The baseline measures are used to prioritize areas of concern for further analysis. Tools for the Analyze phase include multivariable analysis to identify the major family of variation, failure mode and effects analysis (FMEA) to prioritize causes of problematic symptoms, and root-cause analysis using fishbone diagrams or regression analysis. After identifying the main causes of the problem, the Improve phase offers a variety of statistical tools for establishing a better process. Commonly used tools for the Improvement phase include DOE, which consists of tools such as full factorial analysis to determine the effects of the main causes and their interaction. Other experimental techniques include fractional factorial analysis, Taguchi methods, and other designs. One of the key aspects of Six Sigma is the development of breakthrough solutions (i.e., significant improvement through innovative solutions). Development of innovative solutions requires deliberate effort to find a totally different solution. Triz methodology is designed to develop innovative solutions systematically. Once the solution is found, methods such as RSM can be used to optimize results. The Control phase tools are used to sustain the gains realized using

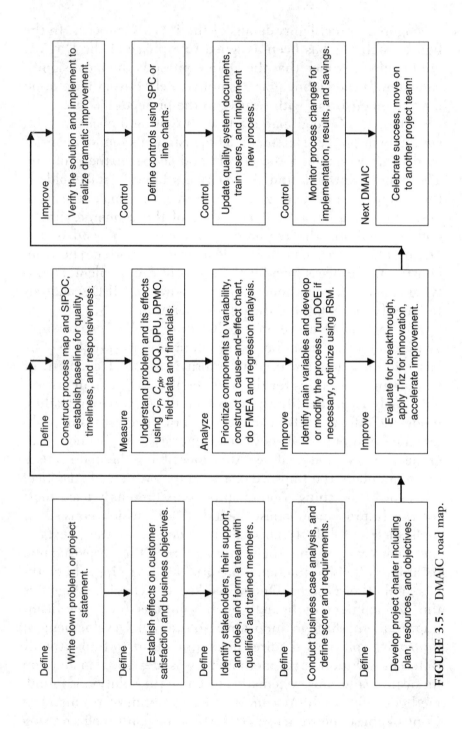

Define
Write down problem or project statement.

Define
Establish effects on customer satisfaction and business objectives.

Define
Identify stakeholders, their support, and roles, and form a team with qualified and trained members.

Define
Conduct business case analysis, and define score and requirements.

Define
Develop project charter including plan, resources, and objectives.

Define
Construct process map and SIPOC, establish baseline for quality, timeliness, and responsiveness.

Measure
Understand problem and its effects using C_p, C_{pk} COQ, DPU, DPMO, field data and financials.

Analyze
Prioritize components to variability, construct a cause-and-effect chart, do FMEA and regression analysis.

Improve
Identify main variables and develop or modify the process, run DOE if necessary, optimize using RSM.

Improve
Evaluate for breakthrough, apply Triz for innovation, accelerate improvement.

Improve
Verify the solution and implement to realize dramatic improvement.

Control
Define controls using SPC or line charts.

Control
Update quality system documents, train users, and implement new process.

Control
Monitor process changes for implementation, results, and savings.

Next DMAIC
Celebrate success, move on to another project team!

FIGURE 3.5. DMAIC road map.

100

DEFINE	MEASURE	ANALYZE	IMPROVE	CONTROL
Kano analysis	Terminology	Cause and effect analysis	Triz	Process thinking
Pareto chart	Normal distribution (Random vs. Assignable)	Multivariable analysis	Comparative experiments (t-test and Z-test)	Precontrol charts
Process mapping	Cost of poor quality	Failure modes and effects analysis	Components search	Control chart concepts
SIPOC	Measurement system analysis	Regression analysis	Full factorial experiments	Internal audits, corrective action, and management review processes
	Performance measurements (C_p, C_{pk}, DPU, DPMO)	Testing of hypothesis	Response surface methodology for optimization	Six Sigma Business Scorecard
		Analysis of variance	Lean thinking and Lean tools	Theory of constraints

FIGURE 3.6. Key Green Belt tools.

the other DMAIC tools. Commonly used tools by Green Belts for all phases of DMAIC are listed in Fig. 3.6. At the successful conclusion of a project, the internal quality system is modified to integrate Six Sigma activities, and the cost and savings are documented. The project team is recognized for successful completion of the project.

SIX SIGMA PROJECT FORMS

Six Sigma has been considered a methodology or a set of tools for the selected few. A lot has been said about the Black Belts. Some Green Belts wonder why they should help Black Belts. It is the Black Belts who get the most benefits from Six Sigma. It is thought that Green Belts help Black Belts; however, this has not proven to be an effective approach. Most Green Belts can be trained to be capable of solving most problems. Six Sigma Project Forms for use by Green Belts are shown in Figs. 3.7 to 3.11, pp. 103–120. These forms correspond to the five phases of DMAIC. These forms guide a practitioner in using the DMAIC methodology and tools to solve most problems with the least effort.

In applying the Six Sigma methodology effectively, the most critical factor is the passionate commitment of the leadership. However, that passionate commitment comes with the correct understanding of the intent of Six Sigma and effective executive support for the Six Sigma initiative. In order to create passionate commitment of the leadership, certain leadership tools and skills must be learned by the executive. These tools and skills are listed in Fig. 3.12. Four important business tools include *employee recognition, process thinking, business scorecard,* and *management review.* Three important Six Sigma tools include *statistical thinking, Six Sigma methodology,* and the *Pareto Principle.* Three improvement tools are *process mapping, cause-and-effect diagrams,* and *rate of improvement.* These 10 tools constitute a minimal set of tools an executive must become familiar with in order to solve a problem or lead employees to solve problems.

Project Title:	Project Leader:
Team Members:	Project Start:

Estimated Project Selection Parameters: Probability of Success (P)_____
Cost (*C*)_____ Time (*T*)_____ Savings (*S*)_____ Project Index (PI) _____

Project Description:

Project Goal and Objectives:

Customer(s):

Customer Critical Requirement:

Project Scope:

Resources Required and Their Source:

FIGURE 3.7a. Six Sigma project management—Define phase.

SIGNIFICANCE OF BREAKTHROUGH

Some believe Six Sigma is very statistical in nature and is suitable only for Black Belts. During the past several years, in the age of DMAIC, Black Belts and Six Sigma have become synonymous. No other level of training is rewarded as much as Black Belt certification. If one achieves Black Belt, corporations recognize it with superior financial value. However, Six Sigma is lot more than just the Black Belt. Six Sigma consists of the intent,

Stakeholder Analysis

Stakeholder ⇨ Support ⇩	Customer	Management	Operations	Quality	Supplier
Passionately committed					
Supportive					
Compliant					
Neutral					
Opposed					
Hostile					
Not needed					

Legend: X = Present level of commitment; O – Required level of commitment

Customer Requirements:

Assumed and unspoken:

Spoken and measurable:

Love to but unspoken:

Critical to Quality (Operation) Requirements:

FIGURE 3.7b.   Six Sigma project management—Define phase. (*Continued*)

methodology, tools, and measurements. For a company to achieve Six Sigma–level performance from a typical performance level (i.e., improvement from the Three Sigma Level to the Six Sigma level), an improvement of almost 20,000 times (66,810/3.4) is needed. Such monumental improvement levels had not been heard of in the corporate world prior to the advent of Six Sigma. Achieving change of such magnitude requires more than incremental and continual improvement. Instead,

Force Field Analysis

Drivers:	Distracters:

Problem Attributes in Order of Significance

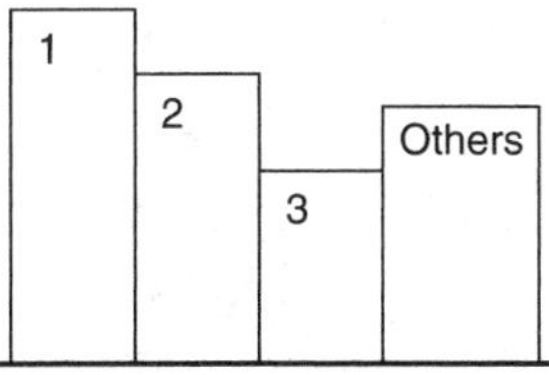

Other Observations:

Measurements

Customer Related (Primary/ External):

Operations Related (Secondary/ Internal):

FIGURE 3.7c.   Six Sigma project management—Define phase. (*Continued*)

one needs dramatic improvement through reengineering and innovation. The challenge facing Green Belts or Black Belts is how to develop breakthrough or innovative solutions. They must learn to look at the process problem differently. This requires certain skills, attitudes, and tools. Black Belts and Green Belts

Process Map

#	Activity	Setup	Wait	Ops.	Store	Move	Insp.	Cycle Time	Quality (L/M/H)	Value Ops. (Y/N)

All times are in minutes.

SIPOC

Supplier (5)	Input (4)	Process (1)	Output (2)	Customer (3)

FIGURE 3.7d. Six Sigma project management—Define phase. (*Continued*)

must find joy and excitement in the battle against the unknown and the struggle to win. Their thoughts must shift constantly, restlessly, between the extreme ranges of the actual and the possible. Black Belts and Green Belts must have the tenacity of pur-

Process Flowchart

Basic Symbols (Additional symbols can be used consistently):

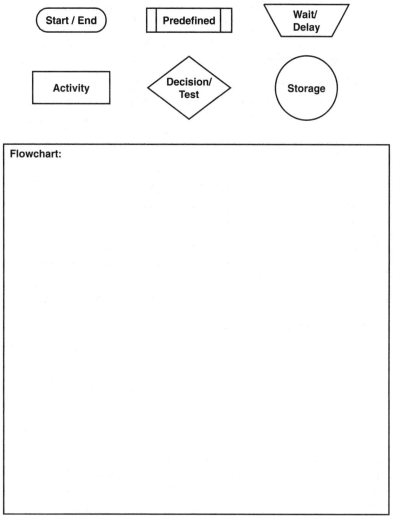

FIGURE 3.7e.   Six Sigma project management—Define phase. (*Continued*)

pose and openness to try any route to the top. One must be a superior opportunist who is looking for ideas to exploit.

Some of the critical skills mentioned by great innovators of past include clarity of thought and capability of expression.

Project Title:	Project Leader:
Team Members:	Project Start:

Estimated Project Selection Parameters: Probability of Success (P)_____
Cost (*C*)_____   Time (*T*)_____   Savings (*S*)_____   Project Index (PI) _____

Project Descriptive Statistics:

Measurements	Average	Standard Deviation	Comments

Cost of Quality

Internal Failure Items	External Failures Items	Appraisal Items	Prevention Items	Summary	Cost
				Internal	
				External	
				Appraisal	
				Prevention	
				Total	

FIGURE 3.8a.   Six Sigma project management—Measure phase.

Black Belts and Green Belts must have the ability to express their ideas and observations correctly in writing. They must be able to focus quickly on the central point in every activity or observation.

In evaluating a process for improvement, one must gather all the information available about the process—setup data, performance data, and data on any changes made to the

Gage R&R Analysis

Critical Gage Description	Part Variation	Operator Variation	Instrument Variation	Total Variation	R&R %

Performance Measures (Use Columns as Appropriate)

Measurement	C_p	C_{pk}	DPU	DPMO	Sigma

FIGURE 3.8b.   Six Sigma project management—Measure phase. (*Continued*)

process. Unacceptable performance is generally attributable to the setup or to excessive variation. Besides, one needs to first understand what has been going on before jumping into a DOE. One must identify related processes and sources of verification of the process performance, and gather observations from various people involved with the process. In gathering

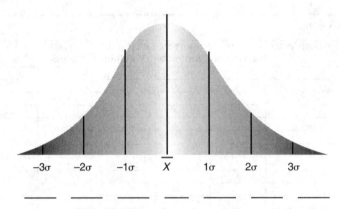

Statistical Depiction of a Characteristic
(Write actual values for a process characteristic)

Unacceptable Key Characteristics and Their Variation	
Characteristics	**Measure of Variation**

Comments:

FIGURE 3.8c.   Six Sigma project management—Measure phase.
(*Continued*)

information about materials, machines, people, and methods, one must ensure that most of the known information about the process is available, implying that the problem space has been captured, that the problem has been bounded. Figure 3.13 shows various sources of data for understanding a process.

Project Title:	Project Leader:
Team Members:	Project Start:

Estimated Project Selection Parameters: Probability of Success (P)_____
Cost (C)_____ Time (T)_____ Savings (S)_____ Project Index (PI) _____

Multivariable Analysis:

Positional Variation (Within piece, or Design related):

Minimum Value: _____ Maximum Value: _____

Positional Variation (Maximum – Minimum)_____

Cyclical Variation (Batch to batch or setup related):

Minimum Value: _____ Maximum Value: _____

Cyclical Variation (Maximum – Minimum) _____

Temporal Variation (Over time, or maintenance related):

Minimum Value: _____ Maximum Value: _____

Temporal Variation (Maximum – Minimum) _____

FIGURE 3.9a.   Six Sigma project management—Analyze phase.

After gathering the information, one can complete the Define phase and establish clear objectives to achieve in solving the problem or achieving the project results. To gain further insight into the process, the process measurements are reviewed and correct measurements are established. The correct measurements reflect the output performance for the basic intent of the process. For an example, in a plating

Failure Mode and Effects Analysis for Identifying Potential Causes								
Process	Potential Failure Mode	Potential Effects of Failure Mode	Sev.	Potential Causes of Failure Mode	Occ.	Current Process Controls	Det.	RPN

FIGURE 3.9b.   Six Sigma project management—Analyze phase. (*Continued*)

process, the basic purpose is to deposit a specified thickness of material. The correct measurement would be the plating thickness, not the number of various cosmetic defects such as bubbles or burnt material. If the part is machined, the correct measurement should reflect the machining process, not just the final dimensions. With the correct measurements, a baseline performance is set up, and quantifiable objectives for

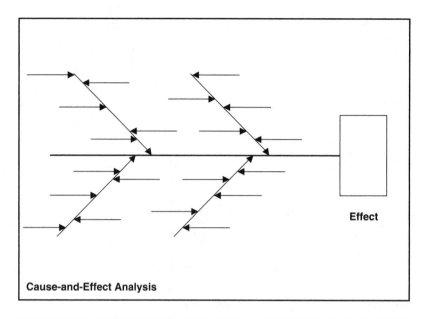

Cause-and-Effect Analysis

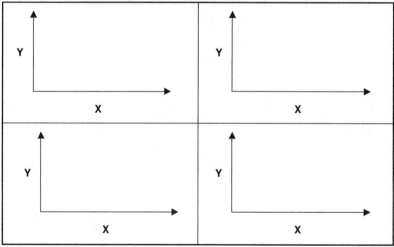

Visual Regression Analysis

FIGURE 3.9c.   Six Sigma project management—Analyze phase. (*Continued*)

improvement are understood. These steps are addressed in the Measure phase of DMAIC.

Now the analysis of the data begins. One can use regression analysis to evaluate various variables based on the available data. The regression analysis between the process output

Hypothesis Testing

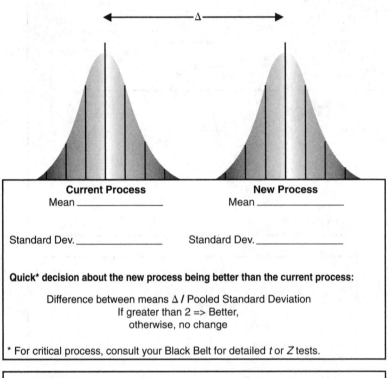

(**FIGURE 3.9d.**   Six Sigma project management—Analyze phase.
(*Continued*)

and various process variables is conducted and correlation coefficient is looked at to prioritize various variables. The quantifiable cause-and-effect analysis is validated with the observations of the cross-functional team members' experience and the operators' observations. Now, with a cross-func-

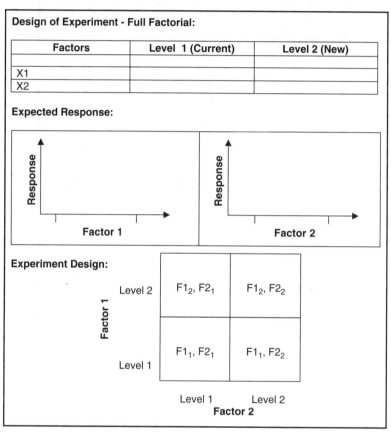

FIGURE 3.10a.   Six Sigma project management—Improve phase.

tional team, the cause-and-effect analysis is performed and all potential variables are listed. The team prioritizes the variables based on their knowledge of the process. It must be emphasized that in-house expertise must be fully exploited, because no experiment can substitute for that in a short time. The objective is to reduce the number of variables to a manageable level and increase convergence between the performance and input variables. Various DMAIC tools can be used

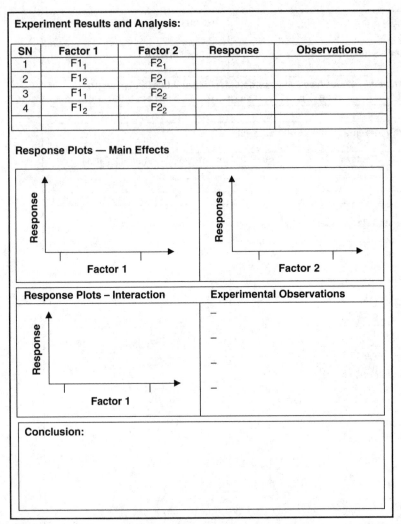

Experiment Results and Analysis:

SN	Factor 1	Factor 2	Response	Observations
1	$F1_1$	$F2_1$		
2	$F1_2$	$F2_1$		
3	$F1_1$	$F2_2$		
4	$F1_2$	$F2_2$		

Response Plots — Main Effects

Response / Factor 1

Response / Factor 2

Response Plots – Interaction   **Experimental Observations**

Response / Factor 1

–
–
–
–

Conclusion:

FIGURE 3.10b. Six Sigma project management—Improve phase. (*Continued*)

to gain a better understanding of the process and the sources of variability.

At this time, team members must start thinking about alternative solutions. Challenge the status quo and seek ideas from all who have worked with the process. Sometimes it helps to examine the basics of the process. The gap between the "as is"

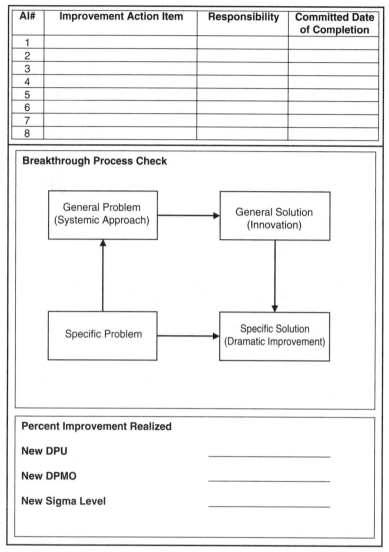

AI#	Improvement Action Item	Responsibility	Committed Date of Completion
1			
2			
3			
4			
5			
6			
7			
8			

Breakthrough Process Check

General Problem (Systemic Approach) → General Solution (Innovation)

Specific Problem → Specific Solution (Dramatic Improvement)

Percent Improvement Realized

New DPU _____

New DPMO _____

New Sigma Level _____

FIGURE 3.10c. Six Sigma project management—Improve phase. (*Continued*)

and the "should be" levels must be investigated. Throughout the project, team members must be committed to performing their assigned tasks on time. Often team members do not complete their action items due to conflicting priorities. This delinquency distracts the team from achieving its objectives. This happens because the organization is not ready to benefit from

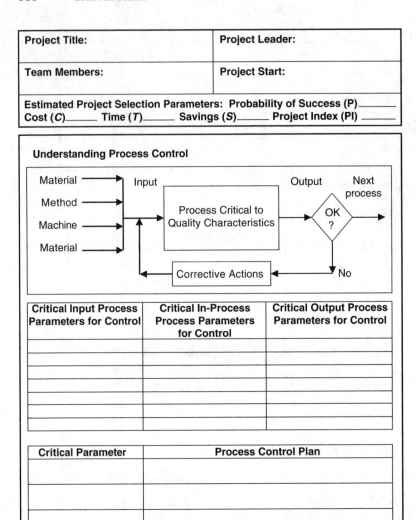

| Project Title: | Project Leader: |
| Team Members: | Project Start: |

Estimated Project Selection Parameters: Probability of Success (P)_____
Cost (*C*)_____ Time (*T*)_____ Savings (*S*)_____ Project Index (PI) _____

Understanding Process Control

Material ⟶ Input

Method ⟶

Machine ⟶ Process Critical to Quality Characteristics → Output → OK? → Next process

Material ⟶

Corrective Actions ← No

Critical Input Process Parameters for Control	Critical In-Process Process Parameters for Control	Critical Output Process Parameters for Control

Critical Parameter	Process Control Plan

FIGURE 3.11a. Six Sigma project management—Control phase.

areas of improvement. Instead, it looks at the job at hand. The psychological readiness of an organization is a critical success factor for breakthrough improvement. Everyone in the organization must understand the significance of the project and its contribution toward corporate success.

While developing a solution, the team must look at the ideal process, the best process, or the optimal process that

X-Bar, R Control Chart

3σ	
2σ	
1σ	
Mean	
-1σ	
-2σ	
-3σ	

UCL

$\overline{R}$

Interpretation Guidelines for Control Charts

Look for random or assignable variation.
Random variation implies no apparent pattern, i.e., maintain bell shaped distribution of all points. In other words, about 2/3 of the points between +/- one sigma lines, about 1/3 between +/- one and two sigma lines, and about 5% between two and three sigma lines.

If there is any pattern in terms of segregation, trend (up or down drift), cycles, stratification or nonrandomness (nonnatural), control chart should identify the assignable cause that is responsible for nonnormality of the data.

If control chart shows no nonrandom patterns, data appears to be normally distributed, therefore, in statistical control.

FIGURE 3.11b. Six Sigma project management—Control phase. (*Continued*)

would produce the best results. One should project one's imagination beyond the current knowledge level. The process of finding an innovative solution is a constant interaction between the hypothesis and the logical expressions it suggests. It is an understanding of what might be and what is in fact the case. It is a systematic way of looking at various combination

Audit of Six Sigma Initiative

Commitment	Results Achieved	Comments
CEO periodically communicates his/her passionate commitment to Six Sigma.		
Six Sigma plan has been shared with employees.		
Six Sigma goals have been clearly communicated and teams established.		
Results of Six Sigma projects are easily noticeable.		
Savings from projects are shared with employees.		
Progress of Six Sigma initiative is being monitored and reported monthly using measurements.		
Our company has improved a lot because of Six Sigma initiative.		

Corrective Actions for Six Sigma Initiative

Concern with Six Sigma	Root Cause	Corrective Action
Too much training		
No success stories		
Six Sigma for a privileged few		
Don't hear about Six Sigma		
Costs too much money		

Management Review Actions

Six Sigma has been added to the agenda.		
Six Sigma progress and savings are reviewed .		
Six Sigma has been incorporated in the quality system.		

FIGURE 3.11c. Six Sigma project management—Control phase. (*Continued*)

Tool/Concept	Type	What (Description)	When (Applicability)
Employee recognition	B	Process of recognizing exceptional improvement activities and employees	To inspire dramatic improvement and employee innovation
Process thinking	B	Understanding business is a collection of processes	Helps understand business processes and how to lead them for improvement
Six Sigma Business Scorecard	B	A corporate performance measurement system balanced for growth and profitability	Learning to achieve improvement in performance and profitability
Management review	B	Role of internal audits, corrective action, and operations management review	Monthly feedback to the management team for necessary adjustment to achieve growth and profitability
Statistical thinking	P	Understanding random and assignable variation	Helps in determining degree of adjustment or type of actions to be taken
Six Sigma overview	P	Understanding the intent, impact, DMAIC, and requirements	Decision making, specifically when committing to Six Sigma
Pareto principle	P	A graphical tool to prioritize commitments based on added value	When deciding about what to work on first
Process mapping	I	Flow charts used to understand information flow, value streams to profitability	Identify disconnections in the business and opportunity for improvement
Cause-and-effect analysis	I	Understanding causative relationship between performance and processes	Identify root cause of problems and remedial actions
Rate of improvement	I	Differences between incremental and dramatic improvement	Achieve dramatic process improvement, reducing waste and achieving profitability

FIGURE 3.12. Key executive tools.

121

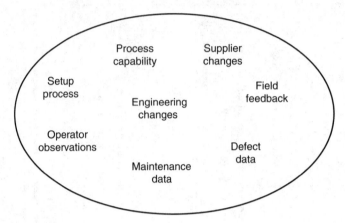

FIGURE 3.13.   Process data sources.

and permutations of variables in the theater of operation. The objective is to find a different recipe that either reduces variability in the process or removes the variables that give rise to the variability of the process. People come up with ideas by association of concepts or ideas, or through a desire for attention or learning. It is a process of understanding the forces acting on the problem. The team challenges itself to find a dramatically better process, or the search for a solution forces the team to look in all directions, beyond currently known levels, to experiences outside work that relate to the process conditions. Just as Einstein reasoned from a painter's fall to a new understanding of gravitational forces, one can bring outside experience to bear in solving problems at work.

Ultimately, the project team comes up with the revolutionary solution to meet the expectations that are set by the corporate Six Sigma initiative.

Triz provides a systematic approach to analyzing the situation at hand and applying various principles of innovation to develop a new solution. For example, if a machine shops uses air to remove the metal chips from parts, another approach could be vacuuming out the chips, totally in an opposite way. Inspired team members equipped with tools and techniques, and challenged to achieve major improvements, do come up with innovative solutions. The idea is that any one of us could become the next Newton, Einstein, Deming, or Juran. One really must believe that the human mind has unlimited potential. Therefore, it is imperative that our minds be challenged by our leadership to outperform ourselves on a continuing basis. When the team has an innovative solution, make a quick check to ensure that it has not been tried before, or ignored or overlooked for some reason, and that it directly correlates with the process output. It is like solving a puzzle. We know when we have solved the puzzle.

After finding the solution, verify that the change in the process produces the desired change in output. Understanding the statistical significance of change is part of statistical thinking (i.e., is the resulting effect a common variation or due to the planned change?).

Practices to Avoid

1. Do not commit to Six Sigma without a proper understanding of its benefits and requirements.
2. Do not work on a project to apply the DMAIC methodology; instead, work on a project to achieve dramatic improvement.
3. Do not join a project team to be visible; instead, join a project team to contribute.

Practices to Promote

1. Intellectual involvement of all employees to generate ideas for improvement and innovation.
2. Recognition of innovative ideas for implementation.

IDENTIFYING THE PROBLEMS—DEFINE PHASE

The most important issue in the successful implementation of any Six Sigma initiative is the identification of the true problems. Correctly identifying the problems justifies the commitment of expensive and scarce resources for maximum returns. One can think of this time investment in project selection or problem identification as the ever-critical *aim* phase of the sequence, "Ready, aim, fire." After all, selecting the correct direction (i.e., the true problem) and making the commensurate effort can lead to the desired results. At the end of this chapter, you should be able to answer the following questions:

What are the customer's critical requirements?

What do customers love to get?

How do we sort out various customer requirements?

How do we prioritize problems, activities, and resources?

Where are the opportunities for improvement and waste in the process?

Who are the players in the theater of operations?

What are the project goals, objectives, and milestones?

Finally, you should be able to understand the various tools used in defining a problem clearly, specifically Kano analysis; Pareto charts; process mapping; the suppliers, input, process, output, and customers (SIPOC) method; stakeholder analysis; and the project charter.

TOOLS

Problem identification and the resulting project selection track closely with the Define phase of the DMAIC model. They include everything from customer identification through needs analysis and clear articulation of the expected improvements and benefits. Following is a list of some of the tools that are widely used to filter out the high-impact problems that have a direct bearing on the bottom line:

- Listening to customers
  Voice of the stakeholders
  Kano model
  Affinity diagram

- Prioritizing for performance
  Pareto analysis
  Force field analysis

- Finding waste
  Process mapping

- Understanding the process and its problems
  SIPOC model
  Project charter

LISTENING TO CUSTOMERS

Motorola initiated its Six Sigma program in the mid-1980s to address competitive challenges by focusing on customers. Upholding the value of the customer was the singular goal that the company rallied around as it began to strengthen its relationships with its customers and drive waste out of its business. This philosophy implied a clear and rapid response to the various needs of internal and external customers as the only way to operate. Mototola's Six Sigma effort paid off, both in the company's successful bid for the 1988 Malcolm Baldridge National Quality Award and in the beginning of a new era in American business, the era of bottom-line business improvements through concerted focus on meeting and exceeding the customer's needs and expectations.

Polaroid, while implementing Six Sigma, remained focused on internal operational metrics rather than external customer-focused metrics. That internal focus brought short-term gains but could not sustain the company's competitive position. Polaroid ran into significant financial problems later.

Conversely, GE, which saved more $12 billion over 5 years, is fully obsessed with its customer focus. It looks at all its processes through the customer's eyes. Jack Welch called this vision "At the Customer for the Customer" (ACFC). This approach ensured that GE evaluated its performance as its customers would. Armed with the customer's view, GE has applied Six Sigma not only to its products but also to the way its business processes deliver services to the customers.

Although the notion that bottom-line business improvement tracks closely with increased customer responsiveness holds true in most cases and is rarely disputed, the translation of this belief into specific, repeatable action is often lost. Many companies that are pursuing a Six Sigma implementation for the first time struggle with the question of properly identifying the customer.

Identifying the customer is a critical step, because if you can identify customers and understand what they consider critical, you can design products and services that meet customer requirements. The objective is to understand the customer's needs, demands, requirements, desires, wants, whims, and, if possible, those things that the customer doesn't even know they want just yet. Customers may be broadly classified in two categories, internal and external.

INTERNAL CUSTOMERS

Internal customers are mainly the employees of the organization. In order to implement the customer-focused culture and remove departmental boundaries, Kaoru Ishikawa coined the term "The next operation as customer." This creates the culture of first understanding and satisfying customer needs within the organization, then extending this awareness to external customers, because of whom the organization exists.

EXTERNAL CUSTOMERS

Simply put, external customers pay the bills. They may vary in size, segmentation, buying power and patterns, but they all matter in a customer-focused business. While various customers' needs may vary in terms of product or service or challenge, most failures to respond to customer needs are not because of special challenges but, instead, because of the simple failure to recognize the customers and seek out their needs and expectations. External customers may be classified as follows:

Primary customers. Those who ultimately use or consume the products and services produced by the organization. They are the end users of the product.

Secondary customers. Those who act as intermediates, sales or distribution channels between the organization and the primary customers.

In analyzing customer requirements, expectations, and feedback, customers can be classified by their relationship with the organization, in the following categories:

- Current customers, happy
- Current customers, unhappy
- Lost customers
- Competitor's customers
- Prospective customers

To serve customers better and grow business with them, you must be proactive in communicating with them and managing their expectations.

VALIDATING CUSTOMER LISTS

This is often the first step in creating a solid path of communication with the customer. Many organizations are short of resources in this area today, and customer contact lists may be outdated or invalid or contain erroneous information. Depending on the source of such information, a customer list can be validated simply through a database cleanup, or it may require substantial case-by-case investigation. In any event, the time spent is worthwhile, as this represents your potential revenue. In most cases, the company should also invest in a long-term solution to prevent mismanagement of this vital information in the future.

VOICE OF THE CUSTOMER ANALYSIS

All companies claim to have a system for customer feedback, but the effectiveness of their systems varies, from the proverbial black hole where customer requests and complaints are logged and forgotten to a highly responsive, real-time feedback structure within the company. As most readers of this book are probably closer to the former than the latter, this should be a priority that is acted upon as quickly as possible. There are two primary approaches to gathering feedback from your customers: an initial survey to gage general customer feelings and an ongoing customer survey. For the company just starting down this path, the former is suggested while the latter is under development. Either should be based on sound principles of brainstorming, focus group studies, or surveys, as follows.

Brainstorming.   Brainstorming is an excellent way of developing many creative solutions to a problem. It works by focusing on a problem and coming up with many radical solutions. Ideas should deliberately be as broad and odd as possible, and should be developed as fast as possible. Brainstorming is a lateral thinking process. During brainstorming sessions, there should be no criticism of ideas. The objective is to explore unlimited or untried possibilities, to generate radical ideas as people give free rein to their creativity. Ideas should only be evaluated after the brainstorming session has finished, when you are exploring solutions. There are two ways of brainstorming:

One-on-one brainstorming. When you brainstorm with one customer at a time, you get many ideas related to that customer. Sometimes, depending on the participating individual's curiosity, this session may be a dead end for ideas. On the other hand, you don't have to worry about many conflicting opinions, as occur in a group setting, and can therefore be more freely creative. However, you might not develop ideas as effectively, as you would lack the broad experience of a group.

Group brainstorming. Group brainstorming can be very effective, as it makes use of the experience and creativity of all members of the group. When individual members reach their limit on an idea, other members' creativity and experience can build the idea to the next stage. Therefore, group brainstorming tends to develop ideas in more depth than individual brainstorming. To run a group brainstorming session effectively, the following steps can be used as a guideline:

1. Define the objectives of brainstorming clearly.
2. Encourage the group members to expand their imagination and to come up with as many ideas as possible, from solidly practical ones to wildly impractical ones.
3. Try to get everyone to contribute and develop ideas.
4. Welcome creative and crazy ideas.
5. Encourage people to develop other people's ideas, or to use other ideas to create new ones.
6. Ensure that no one criticizes or evaluates ideas during the session. This may cause reluctance to put forward ideas.
7. Keep the session focused on the problem.
8. Ensure that no single train of thought is followed for too long.
9. Appoint one person to note down ideas. A good way of doing this is to use a flip chart that can be studied and evaluated after the session.

When possible, participants in the brainstorming process should come from as wide a range of disciplines as possible. This not only brings a broad range of experience to the session but also helps to make it more creative and fun.

Focus group studies. A focus group study is a group activity conducted to generate a rich understanding of participants' experiences, beliefs, attitudes, and needs. In the focus group technique, experts provide ideas and inputs. The experts can either be recruited from within the company (e.g., a team from all layers of management to focus on a general issue) or they can be brought in from outside to provide a fresh set of ideas on the problem.

Focus group studies are a qualitative market research method, often used as a prelude to quantitative market studies such as customer online surveys, phone surveys, and other means of market research. They work best when a limited number of participants, guided by a skillful moderator, have the opportunity to discuss their shared interests within an open environment.

The main point here is the value of getting an in-depth view of the customer's belief and attitude structure, and using this insight to develop a customer-focused strategy. The size of the group affects not only the number and variety of ideas that are exchanged, but also the atmosphere of the session. The normal group size is around seven participants.

The moderator plays a vital role in a focus group. Key attributes of a good moderator include the following:

- Ability to understand the objectives of the focus group
- Ability to design the program, analyze, and synthesize
- Superior listening skills
- Excellent interpersonal skills
- Energetic, enthusiastic, and pleasant demeanor
- Analytical skills
- Expertise in the subject matter

Surveys. A survey is a system for collecting quantitative or qualitative information from the target market to get the feedback on the issues faced by the organization. Surveys involve setting objectives for information collection, designing research, preparing a reliable and valid data collection instrument, administering and scoring the instrument, analyzing data, and reporting the results.

The questions in survey instruments are typically presented in mailed, taped, or self-administered (on paper or a computer) formats or by in-person (face-to-face) or telephone interviews. The following are the features of a good survey.

1. Specific objectives
2. Straightforward questions
3. Sound research design
4. Proper selection of population or sample size
5. Reliable and valid survey instruments
6. Appropriate analysis
7. Accurate reporting of survey results

The following are the steps in constructing a questionnaire for a survey:

1. Decide on the information to be collected.
2. Decide on the target market for collecting the information.
3. Decide on the sample size.
4. Develop the questionnaire in language that is most effective for the target market.
5. Design the survey in an easy to understand and pleasing format.
6. Plan the distribution, return, and follow-up procedures.
7. Write a good introduction on the purpose of the survey.
8. Pilot-test the questionnaire.
9. Determine the questionnaire's technical adequacy.
10. Incorporate feedback from pilot group.
11. Conduct the survey via direct mail, e-mail, telephone, or in person.

12. Analyze the results with respect to the predetermined objectives.

13. Communicate the summary to interested internal or external parties.

While the feedback stream is being established with the customer, the organization must develop an internal system to analyze the feedback and identify specific improvement opportunities. This is where problem identification begins! These improvement opportunities should be further distilled into specific project ideas and worked through to completion. And since the successful implementation of changes based on customer response often requires the involvement of the customer at various stages, this activity can enhance the customer's willingness to do business with the company. Note, however, that customers may not care to understand and explore all of the ways that your product or service can fail them. In other words, you may be required to interpret and exploit meager information from your customers.

Common failure modes of the process of listening to the customer include the following:

1. *Failure to recognize and respond to true external customers.* Hearing customer feedback is one side of a coin. Changing your process or products based on customer feedback is the other. Equally important is to let the customers know about the changes that you have incorporated based on their ideas. It is vital to know your true customers. Sometimes they exist in various forms. For example, in case of a drug manufacturer, there are multiple customers whose roles come into play at different stages:
   - *FDA.* Approve the drug for prescription by doctors.
   - *Doctors.* Prescribe the drug for use by patients.
   - *Drug stores.* Keep the drug in stock for sale to patients.
   - *Patients.* Actually buy the drug to cure an ailment.

2. *Failure to recognize and respond to internal customers.* Internal customers are the key element of the value chain that delivers the value to external customers. The lack of

responsiveness by the participants to the internal customer may eventually result in the loss of the external customer. For example, in the supply chain, lack of timely response to the internal customer's demand may result in the loss of an external customer forever.

3. *Treating all customers alike.*   No doubt all customers are important to the organization, because you can never identify the future Microsofts or Wall-Marts immediately. But at the same time, it is important to keep the business running profitably *today* so as to exist tomorrow. This means taking necessary steps to ensure the necessary care for the organization's most valuable customers, while also nurturing the small customers. In this case, the criteria of value can be the lifetime (potential) profit opportunities with customers.

4. *Not allowing for changes in needs and expectations.*   The customer's needs and expectations keep varying in response to market forces. In today's business environment, *change* is the only known constant. It is vital to conduct environmental scanning on a continuing basis, to ensure that you capture all the little changes in the market faster than your competitors.

KANO MODEL

Noriaki Kano developed a model of the relationship between customer satisfaction and quality. It has become a vital tool of the Six Sigma methodology that helps to focus on the customer's needs and deliver dramatic bottom-line results by delighting the customer. This model classifies customer requirements into three categories:

Basic level of quality in every product (assumed and unspoken).   This represents the most fundamental requirements customers expect from the products they purchase. For example, no one buying a computer would specifically ask whether it includes a hard disk, but its absence would infuriate customers. When you buy a car, you assume it comes with four wheels. In the absence of this basic level, customers will complain loudly.

Expected quality (competitive and spoken).   In this category, customers specify what they want based on their knowledge of the marketplace, product, and service. There are certain quality features that customers value in proportion to their presence. Customers specify features, quantity, or delivery date.

Delighting quality (unexpected and unspoken).   This represents a surprise. Customers generally don't expect this category of features in the product, but when such quality features are present, customers simply love them. For example, Cadillac pioneered a system that keeps the headlights on long enough for the owner to walk safely to the door. Customers always love to get something extra beyond what they have asked for.

Kano analysis is a quality measurement tool used to identify and prioritize customer requirements based on their effect on customer satisfaction. All identified requirements may not be of equal importance to all customers. Kano analysis can help you rank requirements for different customers to determine which have the highest priority.

The Kano process specifically brings the voice of the customer into the design of a new product and service. Kano is a professor at Tokyo Rika University in Japan and an author and consultant who has gained a worldwide reputation for his insight on various aspects of customer-defined quality. Kano recognized the importance in designing any new product or service of maximizing customer satisfaction while avoiding unnecessary extras that add cost but provide little added benefit. The Kano model can be used for the following:

- Identify and understand the customer needs and their classification
- Validate the concepts and the hypothesis about new features being planned
- Perform competitive product analysis
- Determine the new functional requirements

The principles of the Kano process are built around Kano's insights on differing types of customer requirements. His method for sorting the features of a product into various quality categories, based on a questionnaire filled out by customers (Fig. 4.1), offers a process for gaining a deep understanding of customer requirements.

Strengths of the Kano Model

1. *The model brings out the latent thoughts of the customers.* Most of the time, the customers themselves are not very clear about the quality features that they would like to have. But when facilitated by the right process, they can be quite vocal about what would make them feel happy, indifferent, or delighted. These points become too obvious to be ignored.

2. *Customers feel that their satisfaction with some features would be proportional to the feature's presence.* Customer satisfaction is proportional to the magnitude of the feature provided by the organization. The higher the magnitude of the

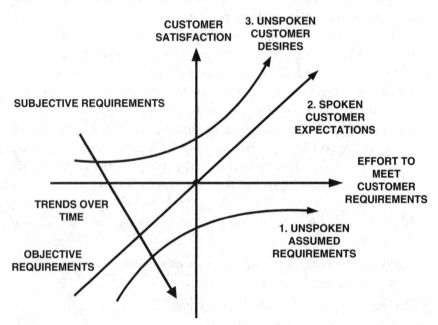

FIGURE 4.1. Kano model.

feature, the greater the customer satisfaction. For example, the greater the hard disk capacity or speed of a computer, the more the customer would love it.

3. *All customer requirements are not linear.* Observe from the model that there are two unique lines—*exciting quality* and *basic quality*. The exciting-quality curve indicates that there are certain features that the customer doesn't normally expect in the product, but would love if provided. This is an excellent opportunity for the organization to seize the competitive advantage by offering as many of these features as is economically feasible. The absence of these features, however, would not hurt the customer. Understanding expectations that the customer would love to have requires building the customer relationship. You really need to understand the customer's feelings, emotions, style, preferences, applications, perceptions, and operating environment.

The basic quality curve indicates features that customers assume are included in the product or service, and any absence or malfunction would dissatisfy them. However, the presence of these features does not get their attention at all, as they are presumed to be essential features that make the product complete. More of these features would not increase customer satisfaction. However, the absence of these unspoken and assumed expectations may cause customer *dissatisfaction.*

4. *The customer's requirements can be explored and classified by someone skilled in customer relations or sales, or by using a well-designed questionnaire.* The questionnaire for the Kano model is unique. For each product feature, it asks questions from two opposing views to classify the true need of the customer.

The following is a typical set of questions:

How would you feel if a particular feature is present?
How would you feel if a particular feature is absent?

To keep the survey process simple, customers could have the following three responses:

• They would love it.

- They don't care.
- They would hate it.

This would mean that for each feature under evaluation, there would be six combinations of responses from the customers (two questions times three responses). These may be summarized as shown in Fig. 4.2.

Considerations for Using the Kano Model

- It is a thoughtful survey that requires a lot of attention to the design of the questionnaire, because the customers may easily become confused by the terms.
- The customers don't necessarily have clear answers to all the questions. Hence, it is important to make the number of questions the minimum necessary to achieve the desired level of understanding of customer requirements.
- The meanings of responses such as "Love it," "Don't care," and "Hate it" must be explained with all synonyms to avoid the misinterpretation related issues.
- Questions need to be validated very carefully with multiple or diverse audiences before administering the survey on a large scale.

Customer's Feedback ⟶		Absence of the Feature		
		Love it	Don't care	Hate it
Presence of the Feature	Love it	Questionable response	Exciting quality	Expected quality
	Don't care	Lack of understanding	Customer is indifferent	Basic quality
	Hate it	Lack of understanding	Lack of understanding	Questionable response

FIGURE 4.2. Combinations of customer responses to a Kano survey.

- The purpose of the questionnaire should be absolutely clear, whether the focus is to test new concepts or validate the hypothesis of existing features.

Once the customer responses are gathered and analyzed, they must be classified for each key feature as follows:

- First, address the basic quality aspects. Customers can not live without these features, and the product or service would be considered incomplete without them.
- Next, address expected quality aspects and provide as many as is economically feasible. This class of requirements is considered to be the measurable. Customer satisfaction is directly proportional to the degree of fulfillment of the requirement.
- Finally, try to provide the most exciting quality features— after all, they are the differentiators for the organization's products and services. Customers are willing to pay a premium of about 5 to 10 percent for these delightful features, perceived to be the *value* features.

Although the model gives powerful insights on the customer's perceptions regarding various categories of customer needs, it is limited by the clarity of the customer's understanding of those features. Customer responses may or may not be consistent over time. This may be due to any of the following reasons:

1. The customer may better understand needs and expectations over a period of time.
2. A competitor may already provide exciting quality features.
3. The customer's expectations may change or become more demanding, therefore requiring continual product or service enhancements and improvements.

This means that the organization must take the time to interpret the Kano survey results but be quick to respond to them, before competitors make the feature an expected quality and raise the bar for exciting quality.

Affinity Diagram

When many ideas are available, an affinity diagram is used to gather facts and ideas to form thought patterns. The affinity diagram is a tool that organizes ideas into groups based on their natural relationships. It is used to discover meaningful groups of ideas in a list of raw ideas.

An affinity diagram is used when the following apply:

1. There is a large volume of data available to address an issue.
2. There is a need to add structure to a large or complicated issue by breaking it down into broad categories.
3. There is a need to encourage a new pattern of thinking.
4. There is a need to come to an agreement on an issue or situation.

The following steps are followed in constructing an affinity diagram:

1. Define the issue or problem to be explored. Start with a clear statement of the problem or goal and provide a time frame.
2. Identify the cross-functional team that will create the affinity diagram. The cross-functional team will ensure that all aspects of the data are taken care of. The members must be knowledgeable about the issue being handled.
3. Have 3- × 5-inch cards or Post-it notes available for writing down ideas, data, facts, or opinions.
4. Collect the cards or notes and spread them out (or stick them) on a flat surface (e.g., a desk or wall).
5. Arrange the groups into similar thought patterns or categories.
6. Develop a main category, idea, or theme for each group. That main category idea becomes the affinity card.

7. Once all cards or notes have finally been placed under a proper affinity card, the diagram can be drawn up.

8. Borders can be drawn around the affinity groups for clarity.

PARETO ANALYSIS

Pareto analysis is used to prioritize issues at hand. It has been observed that any phenomenon in nature demonstrates the "vital few and trivial many" pattern; that is, most of the performance is the result of a few people in an organization, or most of the sales revenue comes from a few customers. J. M. Juran, the quality guru of the twenty-first century, observed that a few people own most of the wealth, and most deaths are attributable to a few diseases or causes. He named this observation after nineteenth-century Italian economist Vilfredo Pareto. The Pareto principle has been observed in other industries, too, as follows:

◘ About 20 percent (that vital few) of the parts may account for 80 percent of the total material cost of a product.

◘ About 20 percent of the customers may provide 80 percent of an organization's total business.

◘ 80 percent of the problems may be caused by 20 percent of the issues.

The so-called 80:20 rule may not be precisely applicable in any situation, but the vital-few-and-trivial-many principle applies in most cases. This implies that all problems, projects, customers, or other things in life at work or at home are not equally important. Nevertheless, when one asks customers to prioritize their requirements, their most likely response is that *everything* is important. Further prodding leads customers to a better understanding of their own needs by forcing them to prioritize various issues. The Pareto principle is a very powerful tool, because all organizations have limited resources with lots of opportunities for improvement. While listening to customers or gathering information to identify potential projects, one can use the Pareto principle to prioritize potential projects based on their effect. By conducting the Pareto analysis, one can focus on a few vital problems for a high return on investment, instead of working on many trivial projects with an insignificant return on investment. Once a few major projects are identified, the appropriate resources, in training, equipment, or support, can be allocated to solve the problem. In the Define phase of the DMAIC method, Pareto analysis can be used to prioritize problems or projects. In solving problems, the Pareto chart can be used to prioritize the effects of different variables, in order to determine which variables are significant.

The Pareto chart is used to prioritize a data set through a graphical depiction of the data. When process data are collected, Pareto analysis is used to determine which problem to work on first. As mentioned earlier, working on the right problem, and defining that problem correctly, are critical aspects of problem solving and achieving dramatic improvement. Therefore, the Pareto analysis can be used to do the following:

1. Analyze a problem with a new perspective.
2. Focus attention on problems in priority order.
3. Compare data changes during a different time period.
4. Provide a basis for the construction of a cumulative line.

There are many software tools that can perform a Pareto analysis once the data are entered. The following are the steps in constructing a Pareto chart:

1. Determine the classifications (Pareto categories) for the data analysis.
2. Select a time interval for analysis. This interval should be representative of typical performance.
3. Determine the total occurrences (i.e., cost, defect counts, etc.) for each category from the data collected. If several categories occur infrequently, they may be combined and categorized as "others."
4. Rank the occurrences from most frequent to least frequent.

5. Determine the grand total and calculate the percentage contribution of each category.

6. In the next column, calculate the cumulative percentage for each category. The cumulative percentage helps to see the vital few.

7. Plot the data on the bar graph, with the left vertical axis as the frequency of occurrence and the right vertical axis as the percentage contribution. Look for the 80:20 break point on the cumulative percentage graph. This can be easily identified by the sharp change in the shape of the graph.

For example, the ABC company receives a number of complaints about its XYZ product, and the management wants to understand which complaint to resolve first. The customer complaint data for an entire year is shown in Fig. 4.3.

In order to perform a Pareto analysis, the data are sorted by the frequency of occurrence of the problem. Then, the percentage contribution and cumulative frequency of occurrence are determined, as shown in Fig. 4.4. The sorted frequency distribution, cumulative frequency of occurrence, and Pareto charts are shown in Figs. 4.4 to 4.6.

From the cumulative line, one can observe that the problem types E, D, and A contribute 80 percent of the total problems

TYPE OF PROBLEM	FREQUENCY OF OCCURRENCE
A	170
B	10
C	60
D	200
E	230
F	5
G	30
H	25
I	20

FIGURE 4.3. Problem frequency distribution over 1 year.

TYPE OF PROBLEM	FREQUENCY OF PROBLEM	PERCENTAGE CONTRIBUTION	CUMULATIVE FREQUENCY	CUMULATIVE PERCENTAGE CONTRIBUTION
E	230	30.7	230	30.7
D	200	26.7	430	57.3
A	170	22.7	600	80.0
C	60	8.0	660	88.0
G	30	4.0	690	92.0
H	25	3.3	715	95.3
I	20	2.7	735	98.0
B	10	1.3	745	99.3
F	5	0.7	750	100

FIGURE 4.4.   Problems sorted by frequency of occurrence.

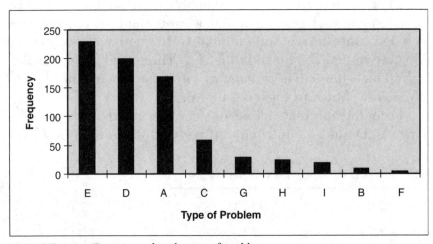

FIGURE 4.5.   Frequency distribution of problems.

faced by the organization. Effort needs to be directed to address these problems first.

WHEN THE PARETO CHART DOES NOT CONFORM TO 80:20

The preceding example covers most real-life situations. However, there are situations when the data distribution is not

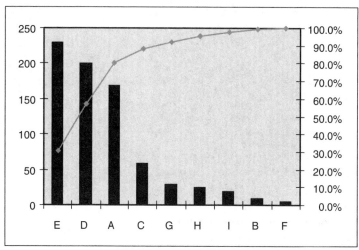

FIGURE 4.6. Cumulative frequency distribution of problems.

distributed precisely according to the Pareto principle. Instead, the data may be more widely and approximately uniformly distributed, as shown in Fig. 4.7. The frequency distribution, Pareto chart, and Pareto chart with a cumulative line are shown in Figs. 4.8 to 4.10.

Note that the frequency distribution is almost flat, and no single bar is significantly different from the others. Under such circumstances, a weighted Pareto analysis can be used to prioritize problems.

WEIGHTED PARETO ANALYSIS

When the frequency distribution does not conform to the Pareto principle, the data are transformed to reflect the effect of the problem by assigning a financial value to each problem. In other words, a different dimension is investigated for prioritization. The data for costs associated with the problem are shown in Fig. 4.11. The total cost of each problem is shown in Fig. 4.12, and the Pareto charts are shown in Figs. 4.13 and 4.14.

Analysis based on the effect of the problem rather than the frequency of distribution clearly identifies the significant problems. One can also see an inflection, or the break point, in the cumulative line. Now the top three problems account for about 85 percent of the total cost effect on the company.

TYPE OF PROBLEM	FREQUENCY OVER PAST YEAR
A	125
B	85
C	95
D	100
E	150
F	75
G	60
H	55
I	120

FIGURE 4.7. Problem frequency distribution over 1 year.

TYPES OF PROBLEM	FREQUENCY OF PROBLEM	PERCENTAGE CONTRIBUTION	CUMULATIVE FREQUENCY	CUMULATIVE PERCENTAGE CONTRIBUTION
E	150	17.3	150	17.3
A	125	14.5	275	31.8
I	120	13.9	395	45.7
D	100	11.6	495	57.2
B	95	11.0	590	68.2
C	85	9.8	675	78.0
F	75	8.7	750	86.7
G	60	6.9	810	93.6
H	55	6.4	865	100.0

FIGURE 4.8. Problems sorted by frequency.

PARETO (PROJECT) PRIORITY INDEX

This concept was developed by Juran and Gryna in 1993. The Pareto priority index is a tool to prioritize projects. It is called the project priority index (PPI) in Six Sigma methodology. The PPI is calculated according to the following formula:

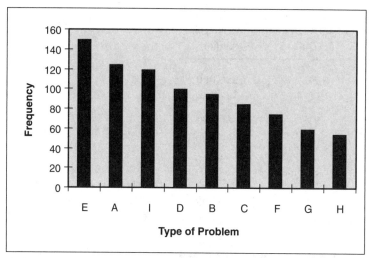

FIGURE 4.9.   Frequency distribution of problems.

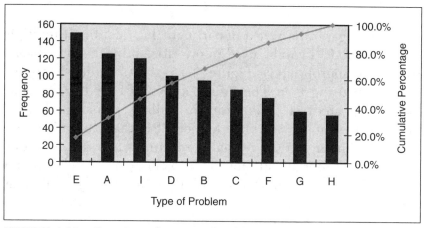

FIGURE 4.10.   Cumulative frequency distribution of problems.

$$\text{PPI} = \frac{S \times P}{C \times T}$$

where S = potential saving generated by the project
P = probability of project success
C = cost of executing the project
T = time to complete the project

TYPE OF PROBLEM	FREQUENCY OF OCCURRENCE OVER 1 YEAR	COST PER OCCURRENCE
A	125	$30,000
B	85	10,000
C	95	80,000
D	100	90,000
E	150	5,000
F	75	4,000
G	60	3,000
H	55	10,000
I	120	5,000

FIGURE 4.11. Frequency distribution and cost data.

If an organization or a department has identified multiple projects, the PPI can be used to prioritize projects and allocate resources appropriately. Figure 4.15 shows multiple project evaluation using PPI. The figure shows PPI for five projects. From the analysis, reduction in billing errors and yield improvement appear to have higher PPIs than the other projects. Depending on the available resources and the corporate need, an appropriate number of projects can be selected for improvement. It is quite possible that the management may decide to work on the ISO 9001 project before attacking the billing errors and yield-improvement projects, because of customer demands. It is also possible that all three projects may be selected for improvement.

FORCE FIELD ANALYSIS

Force field analysis is a technique based on the premise that change occurs after some struggle between *restrainers* (forces that impede change) and *drivers* (forces that facilitate the change). Exploiting the drivers and suppressing the restrainers can accelerate the desired change. Using force field analysis

Type of Problem	Frequency (Past Year)	Cost per Occurrence	Total Cost	Percentage	Cumulative Cost	Cumulative Percentage
D	100	90	9000	39.3	9000	39.3
C	85	80	6800	29.7	15800	69.1
A	125	30	3750	16.4	19550	85.4
B	95	10	950	4.2	20500	89.6
E	150	5	750	3.3	21250	92.9
I	120	5	600	2.6	21850	95.5
H	55	10	550	2.4	22400	97.9
F	75	4	300	1.3	22700	99.2
G	60	3	180	0.8	22880	100.0

FIGURE 4.12. Problems sorted by total cost.

149

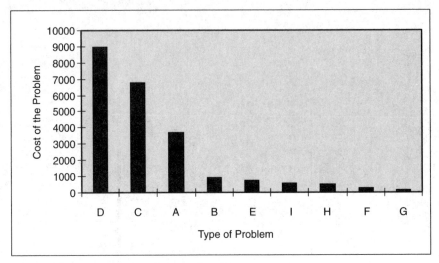

FIGURE 4.13. Cost distribution of problems.

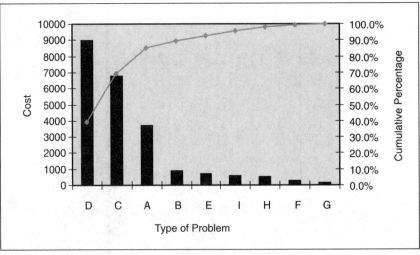

FIGURE 4.14. Cumulative cost distribution of problems.

forces the company to identify all facets of a desired change. By using force field analysis, one can learn which course of action will be the best one to implement, with the most driving forces and the least resistant forces. Also, once the drivers and restrainers are known, it will be easier to devise a plan that will manage the forces that restrain progress, and concurrently exploit the forces that propel in the desired direction.

PROJECT	ANNUALIZED SAVINGS	PROBABILITY OF SUCCESS	TOTAL COST	TIME FOR COMPLETION, YEAR	PPI
Inventory reduction	$250,000	.8	$100,000	0.5	4.00
Yield improvement	2,000,000	.6	250,000	0.33	14.55
ISO 9001 registration	5,000,000	.9	500,000	1.5	6.00
Field failure reduction	200,000	.5	100,000	0.75	1.33
Billing error reduction	100,000	.75	10,000	0.25	30.00

FIGURE 4.15. Pareto priority index for multiple projects.

Force field analysis is widely used in change management and can be used to help understand most change processes in organizations. As shown in Fig. 4.16, to implement a change in the current process and achieve the desired state, the drivers must overcome the restrainers and move toward that desired state.

Force field analysis is mainly used in the early stages of planning for the following purposes.

- To understand and strengthen the driving forces for change (Example: What things are driving the organization toward the desired change?)
- To identify the obstacles or restraining forces to change (Example: What is restraining the organization from achieving the desired results?) and manage them toward the desired state
- To improve the ratio of driving to restraining forces to implement the change

In force field analysis, *change* is characterized as a state of imbalance between driving forces (e.g., changing markets, new technology) and restraining forces (e.g., individuals' fear of fail-

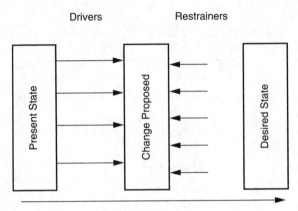

FIGURE 4.16. Force field analysis.

ure, organizational inertia). To implement the Six Sigma initiative, the following analysis could be helpful:

- Define the new state or the process.
- Explain the process to the group of about six people.
- Create a chart with two headings: *driving forces* and *forces of resistance*.
- Gain consensus on the current state and the desired state.
- Brainstorm the driving and restraining forces.
- Prioritize driving and restraining forces for appropriate action.
- Develop an action plan to exploit drivers and manage restrainers.

For example, say someone is living in an apartment and would like to buy a house. Figure 4.17 shows the driving and restraining forces that will facilitate the decision to buy a house, or change the present state to a future state. Similarly, if one needs to improve a process, driving forces could be customer expectations, waste reduction, and reduction in excessive checking, and restraining forces would include the process owner's resistance to change, conflicting priorities, or lack of resources.

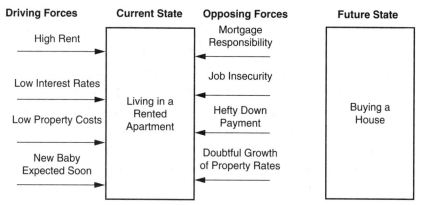

FIGURE 4.17.   Example of force field analysis.

PROCESS MAPPING

The purpose of process mapping is to gain a better understanding of a process, establish a baseline, and identify disconnection in the process. To improve a process quickly, one can map a process, identify disconnection, and fix them. Mapping a process is a cross-functional activity for accuracy and adequacy.

Advantages of Process Mapping

1. Process maps graphically illustrate and convey the necessary details of the process. The functional boundaries otherwise limit the appreciation of a complete operation. The process map identifies process interfaces, minimizing disconnections between processes.

2. Process maps are used to identify opportunities for improvement. For example, one may wonder in the preceding example that out of the total cycle time of 81 minutes of the cutting operation, only 8 minutes (less than 10%) is required for the actual cutting, and the rest is all non-value-added activities.

3. Process mapping increases the participation of the members who work on different aspects of the process. The process map becomes a good training tool for new and existing employees.

4. The basic activity of developing a process map provides a practical insight as understanding is reconciled, waste is identified, and improvement opportunities become evident.

5. The process map highlights areas of suboptimal performance. When a cross-functional team reviews a process map, the team is practically forced to look for the ways and means to reduce the time spent in non-value-adding activities.

6. In practicing the DMAIC methodology, the process map helps in defining the problem better and assists in establishing a baseline performance.

A *process* can be defined as a series of tasks that transforms a set of inputs into desired outputs. The process maps consists of one or more progressively connected activities to achieve the end results. The more steps, the more opportunities for improvement exist in a process. A process requires material, method, tools, and people. A set of activities without established and documented procedures becomes a practice that is dependent on the people performing those tasks. Practices are more prone to variation and problems. However, developing a process map or documented procedure capturing those effective practices defines a process. The following equation depicts the relationship between process and practice:

$$Process = practice + procedures\ (process\ map)$$

A well-defined process is repeatable and reproducible. Following are the typical characteristics of a process:

- Every business function can be considered as a process.
- Processes have a beginning and an end state.
- Processes are coordinated activities that involve people, procedures, material, technology, and infrastructure.
- Processes have associated costs.
- A collection of processes make up the business.
- A business's strength is its weakest process.

▪ Processes must be designed to produce a product or service to meet customer requirements.

A process map is a graphical and logical representation of activities, showing the sequence of tasks using standard flow-charting symbols. It provides a picture of how the work flows through the company. It creates the vocabulary to help people discuss process improvement. It identifies a series of steps or events that take place over a specified period of time.

A process always starts with a triggering event and flows to an endpoint. This trigger can be internal or external. It could be a customer placing an order or the decision to address a customer complaint. Between the triggering event and the endpoint is a series of inputs and outputs, with a defined sequence and defined functional responsibilities. More often than not, these responsibilities are cross-functional in nature.

Process mapping is the best way to visually and clearly articulate how work flows through your organization. The usual approach to mapping a process is to construct an "as-is" map to identify the current status of a process; analyze the map for its performance in terms of quality, timeliness, and cost to identify disconnections; and make changes to the map. The analysis is used to identify process steps that are the potential cause of bottlenecks, delays, barriers, and errors—and to create a map of the reengineered process to aid in selling identified process improvements.

Each process has inputs, in-process activities, outputs, verification of the outputs, and necessary corrective activities to maintain the equilibrium. Input is actually what is fed into the system for processing to produce something of value for the customer. The examples of input are material, labor, skill, equipment, and methods. The in-process activities also include controls at critical activities to ensure that those activities are performed as desired. Otherwise, the process output would be adversely affected. Examples of intermediate processes could be manual or automated processes such as assembly, annealing, testing, or packaging. The output is the end result of the process. The output is sent directly to the internal or external customer. The internal customer could be the next process, including shipping, or even the outsourced process. The external customer is the ultimate party that has requested and pays for the product or service. Examples of outputs include finished goods or services, and semifinished goods or services, which would become the input for some other process. One must be concerned about the undesirable outputs such as scrap, rework, and pollution.

Measures of effectiveness or verification are critical aspects of the process. In most cases, people are busy doing the work. However, they have limited information about how well the work is being done. Good processes have built-in verification activities to ensure that only good outputs are sent on to the next process. Nonconformities are identified through verification, analyzed, and corrected appropriately.

FLOWCHART SYMBOLS

There are industrial standards for constructing process maps. However, for internal process maps, a standard set of symbols can be used. Inconsistent use of flowchart symbols leads to miscommunication of process steps and creates additional opportunities for error. The flowchart symbols standardized by the American National Standards Institute (ANSI) and the symbols that are frequently used are shown in Fig. 4.18.

Constructing a Process Map

1. Select a process to be mapped.
2. Define the process by clearly identifying the boundaries, and stating its purpose and scope.
3. List all the activities that occur in the process.
4. Sequence the activities, and understand the process inputs and outputs.
5. Construct the flow chart using established symbols.
6. Connect various process steps.
7. Identify areas where verification and repair activities are performed.

Symbol	Description
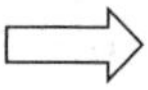 ⇨	Describes the movement of the worker, the material, or the equipment
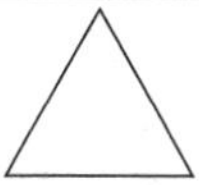 △	Describes the storage; where material is received into or issued from a store, or an item is retained for reference purposes
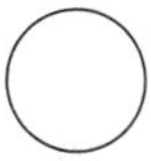 ○	Describes the operation: a main step, where the part, material, or product is usually modified or changed
□	Describes the inspection: a check for quality or quantity
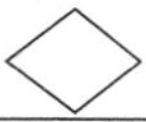 ◇	Describes the decision point: if okay, then one route; else, the other route
 D	Describes the delay: the wait to perform the next step
 ⬡	Describes the preparation: the setup of the machine

FIGURE 4.18.   Flowchart symbols.

8. Capture process Q, T, and C performance at critical steps.

Common Pitfalls of Process Mapping

1. Spending too much time in constructing the process map

2. Making too detailed a description of the process

3. Overlooking the verification activities

4. Ignoring the purpose of the process

5. Missing the repair or rework activities

6. Not specifying control actions

The process map can be generated at the business level, department level, or process level. The business-level map is at a high level to depict the information flow of the major business processes. The business-level process map crosses the departmental boundaries, and shows the information flow across the departments. The process-level map identifies the details of a process. An example of a manufacturing process in which a steel part is cut and sent to the next operation, milling, is shown in Fig. 4.19. The map displays the flow of information and can act as a visual tool for training, identify opportunities for improvement, and establish the current state of the process.

SIPOC

SIPOC is an abbreviation for *supplier, input, process, output,* and *customer.* The SIPOC method is an excellent tool to expand the understanding of the information and its sources in a process flow. Understanding the various players and their contribution to a process can help identify variables in the process, critical aspects of the process, measures of effectiveness, and intermediate levels of supplies and deliverables. The SIPOC model answers the following questions:

1. Who is the work done for?
2. What does the process produce?
3. How is the work done?
4. What is needed to do the work?
5. Who fulfills the needs?

An outline of a SIPOC diagram is shown in Fig. 4.20. If the process $P1$ represents shipping a container of paint to a cus-

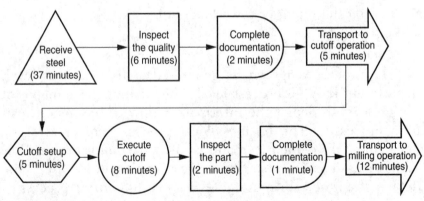

FIGURE 4.19.   Process map of the cutting operation.

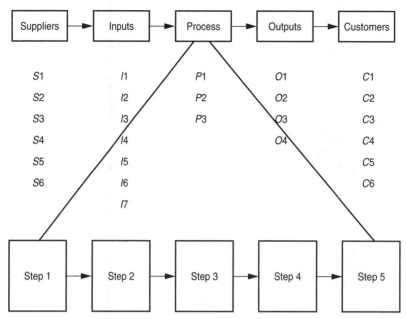

FIGURE 4.20.   SIPOC diagram outline.

tomer, the input would be the container, label, packing slip, and shipping document. The suppliers would include the paint supplier, the printer for the label and packing slip, and the shipping department for shipping documents. The output would be the packaged paint ready to ship to the customer, the auto body shop.

The SIPOC diagram can come when defining or diagnosing the problem. The customer may see the problem as paint not being available for production. However, SIPOC can reveal that the problem is with a certain process, with a certain output, and with certain inputs. Note from this simple example the power of the SIPOC model. A sample SIPOC table for committing to a Six Sigma initiative is shown in Fig. 4.21.

CONSTRUCTING A SIPOC TABLE

In constructing a SIPOC, one can take different approaches. Some people start completing input, suppliers, process, outputs, and customers in any order, or even concurrently. Another approach is by element, selecting one of the five ele-

SUPPLIER	INPUT	PROCESS	OUTPUT	CUSTOMER
▪ Right training provider	▪ Leadership education	Commit to Six Sigma	▪ Commitment to corporate growth and profitability through Six Sigma	All stakeholders:
▪ Performance information	▪ Drivers for Six Sigma Market position		▪ Revised vision and direction	▪ Customers
▪ Market intelligence			▪ Expected financial gains	▪ Employees
▪ Decision to perform business opportunity analysis	▪ Business opportunity analysis			▪ Owners
▪ Candidates to drive improvement	▪ Qualified resources			▪ Management
▪ Resources for competitive information	▪ Competitive assessment and awareness			▪ Shareholders
▪ Corporate performance system	▪ Corporate performance			

FIGURE 4.21. Sample SIPOC table.

ments of SIPOC, completing it, and moving on to the next one. Another methodical approach is to use the process map, insert it in the SIPOC, and complete the output and customer columns. Outputs of each process are listed in the output column, and their destinations are listed in the customer column. After completing the outputs and customers, inputs are identified and listed in the input column. Input sources are listed in the supplier column. When constructing a SIPOC table, one needs to be methodical, patient, and thorough. Using a cross-functional team would expedite the completion of the SIPOC table.

UNDERSTANDING STAKEHOLDERS

When completing the SIPOC table, you identify suppliers and customers. These are critical stakeholders in the success of a project or a Six Sigma initiative. Stakeholders are anyone who has an interest in the performance of an organization. The right stakeholders and their critical needs may also help in the identification of opportunities for the company. A list of stakeholders may include the following:

- Customers of the organization
- Suppliers to the organization
- Employees of the organization
- Producers of the substitute products and services
- Stockholders of the organization
- Lenders to the organization
- Regulators of the business environment
- Community in which the organization operates

Companies that successfully implement Six Sigma address the needs of as many stakeholders as possible. Different stakeholders may have different needs. For example, customers are interested in receiving good parts on time at a reasonable price. Customers also want better, faster, and cheaper products or services continually. Employees would like to have a good working environment, competitive compensation, more tools to make their jobs easier, and the opportunity to grow. Suppliers want more business, better prices, and less interference by the customer. Stockholders want value for their money, that is, increasing market value and dividends. Similarly, regulators have their requirements for compliance, and the community has its expectations for contributions from the corporation.

The purpose of the stakeholder analysis is to understand the stakeholders' needs, their interest in the Six Sigma initiative, or their interest in a specific project. Besides, one needs to know which of their priorities conflict with the Six Sigma initiative, their level of commitment to the Six Sigma initiative or lack of it, and their roles in achieving the desired results. One must ensure that all stakeholders support the Six Sigma initiative, and that appropriate resources are available.

A highly effective Six Sigma project will bring major changes to a system or to the entire company. The change can affect various groups, individuals, or stakeholders inside and outside the company. This makes it necessary to identify the correct stakeholders and address their concerns to obtain their support for successful completion of the project. Stakeholders can be classified as follows, based on their effect on the project outcome:

- *Key stakeholders* are those who are capable of significantly influencing the success of the project but may not be immediately affected by the project activities in the short term, for example, the C-level leadership, and even customers. These stakeholders have a controlling and partially participating interest. These stakeholders must be fully committed to the Six Sigma initiative and projects.

- *Primary stakeholders* are those who are immediately affected by the project activities, for example, the project team members. They may have a noncontrolling and active participation interest. They may also include customers and suppliers. These stakeholders must be committed to the completion of the assigned action items.

- *Secondary stakeholders* are those who are indirectly affected by the project activities and have a noncontrolling and inactive participation interest, for example, some customers, suppliers, or shareholders. These stakeholders must be interested in the welfare of the project.

While solving a problem, the stakeholders must be heard, and their requirements and expectations must be taken care of. One must assess stakeholders' expected support and level of commitment. If there is any gap between the required support and the level of commitment, one must ensure that this gap is eliminated or reduced significantly. The gap must be continually assessed and minimized.

One can view the stakeholder analysis through four *Is*. The four *Is* are *identification, involvement, interest,* and *influence.* The identification of stakeholders early in the project definition phase allows the project leader to ensure that stakeholders' support is available. Once the stakeholders are identified, one needs to assess their involvement in the project. Are they key, primary, or secondary stakeholders? Then, their interest in the project must be understood. Is the project related to something they like to do, learn, or support? If they are interested in the project, stakeholders' commitment improves as the project progresses. If they are not interested in the project, their commitment dwindles over time. Besides their personal involvement, one may also want to capitalize on their influence level for the success of the project. Some stakeholders may not be able to contribute to the success of the project personally; however, they may find other individuals to support it. The four *Is* are as follows:

$I1$ Identification of stakeholders

$I2$ Involvement required

$I3$ Interest at present

$I4$ Influence on project outcome

Besides, being technical, capable, or Black Belt certified may not guarantee the success of the project. Being resourceful, knowing stakeholders, and using their influence is an equally important aspect of successful problem solving. For example, a Black Belt wanted some Gage R&R analysis. He quickly prepared a long document, gave it to three operators, and asked them to complete analysis. He told the operators that he wanted all of this completed by the end of two business days later. As soon as he walked away, people started wondering what they were supposed to do. Looking at all the numbers and statistical equations, the operators had no idea what to do, and they were so busy they had no time to complete the analysis in the time frame the Black Belt required. They just did not do it. On the other hand, the Black Belt was thinking he had done a wonderful job in getting the analysis done. When he returned, of course, he was disappointed. Then he asked what he needed to do to get the project done in time. The operators told him that their boss told them to put the Gage R&R analysis aside for open schedule.

What could the Black Belt have done differently? He could have asked the operators for their input in designing the plan to make it compatible with their schedule, he could have talked to their supervisors about their priorities, and he could have met with their leadership to gain consensus for doing the study. Knowing the informal as well as the formal structure for getting things done is just as important as knowing how to do the project. Good project leadership involves as many stakeholders as possible, if for no reason other than to find the skeptics, because they can derail the project.

Following is a summary of the process for conducting a stakeholder analysis:

1. Identify all the potential stakeholders.
2. Identify the level of involvement in the project desired for them and categorize them as key, primary, and secondary.
3. Identify their current level of interest in the project.
4. Assess their interest level.
5. Identify their positive and negative influence on the project.
6. Assess their universe of influence.
7. Develop a strategy to involve all stakeholders.

8. Pursue the stakeholders with the maximum gap between the requirements and their commitment. While the favorable stakeholders will, by definition, support the project, the unfavorable ones need to be persuaded to do so.

9. The strategy to persuade questionable stakeholders may include the following points:
   - Understand the reasons for their skepticism and address their concerns.
   - Gain the help of the favorable forces in influencing the unfavorable ones.
   - Identify additional stakeholders who may favorably affect the unfavorable ones.

STAKEHOLDER ANALYSIS

The stakeholder analysis can be performed using the template shown in Fig. 4.22. Most of the project stakeholders should fall under one of the eight categories identified. Once the position of the stakeholder is defined with a reasonable degree of confidence, the following actions can be initiated:

1. Move the stakeholders from negative influence status to positive influence status by aligning their involvement and interest levels.

2. Increase their current low interest level to the desired high interest level commensurate with their involvement requirements.

The stakeholder with low involvement, high interest, and positive influence can help with the low-involvement, negative-influence, high-interest stakeholders.

For example, a Black Belt initiates a project that evaluates alternative materials at the shop floor to improve the process output. The supervisor and operators are currently noncooperative, while the management team believes that the experiments must be conducted. The present material supplier is also threatened by the potential loss of business. In this case, the

STAKEHOLDER	INVOLVEMENT	INTEREST	INFLUENCE
Primary/key	High	High	Positive
Primary/key	High	Low	Positive
Secondary	Low	High	Positive
Secondary	Low	Low	Positive
Primary/key	High	High	Negative
Primary/key	High	Low	Negative
Secondary	Low	High	Negative
Secondary	Low	Low	Negative

FIGURE 4.22.   Stakeholder analysis table.

stakeholder template (shown in Fig. 4.23) leads to the following actions:

1. The Black Belt needs to understand the concerns of the plant manager and the materials manager and increase their interest level to high, in line with the involvement level desired for the success of the project.
2. According to the influence patterns shown in Fig. 4.24, high-interest positive-influence stakeholders can influence any negative-influence stakeholder.

PROJECT CHARTER

The project charter is the deliverable of the Define phase. It summarizes the project definition by stating the project objectives, identifying the participants, developing a schedule for project completion, and establishing milestones (quantifiable and measurable goals). After documenting these aspects of the project definition, one can commence the project execution with full visibility. A project charter is a written road map that does the following:

1. Describes the business case, including cost-benefit analysis.
2. Defines the problem to be addressed by the project.

Stakeholder	Involvement	Interest	Influence
Black Belt	High	High	Positive
Plant manager/ materials manager	High	Low	Positive
HR manager	Low	High	Positive
Accounts manager	Low	Low	Positive
Supplier	High	High	Negative
Production supervisor/ operators	High	Low	Negative
Inspectors	Low	High	Negative
Maintenance supervisor	Low	Low	Negative

FIGURE 4.23.   Sample stakeholder analysis table.

Who should influence	Whom
Plant manager	Production supervisor and operators
Materials manager	Supplier
HR manager	Maintenance supervisor and inspectors

FIGURE 4.24.   Who should influence whom.

3. Specifies the project scope.

4. Declares the goal statement.

5. Defines the roles of the team members.

6. Establishes the timeline, milestones, and key deliverables.

7. Identifies resources and other requirements.

The project charter is referred to throughout the project, ensuring the desired progress. One should control the tendency toward project creep. The project charter should not change significantly throughout the project life cycle, unless the project changes. It is created at the beginning of the project, is approved by the key stakeholders, and is available for reference throughout the project life cycle. It is a single, consolidated source of information about the project in terms of initiation and planning, and provides information about project scope, objectives, deliverables, risks, and issues. It also lays the foundation for how the project will be structured and managed in terms of change control, oversight, risks, and issues. While developing the project charter, the project leader must ensure that:

1. The team goals are aligned with the organizational goals.
2. The team obtains full management commitment.
3. The team is aware of the project goals and boundaries.
4. The team can remain focused on the defined goals.

BUSINESS CASE

Developing a business case is a critical step in launching a Six Sigma project. The business case justifies the project and facilitates allocation of the necessary resources. Whether the project is the companywide implementation of a Six Sigma initiative or the launch of a Six Sigma project, the leadership and associates must see benefits or effects on the bottom line. This is the reason for undertaking the project. The business case communicates the reason for committing the organization's expensive resources. It states the organizational goals and explains how the project goals will support business objectives. It establishes the performance metrics that will be affected positively by the project results, making it easy to verify completion of the project and achievement of the planned savings.

An important aspect of preparing the business case is the cost-benefit and return on investment (ROI) analysis. ROI is the ratio of the net gain from a proposed project to its total costs. ROI is a great tool for showing how much each dollar spent will yield in returns. An accurate ROI analysis includes both the tangible and intangible benefits.

When computing the cost, it is important to include known, hidden, direct, and indirect costs. Any cost reduction must ensure cost *elimination*, not cost *transfer* to some other player in the value chain. The cost and benefit assumptions must be documented and questioned for validity as necessary.

There are several methods for calculating the value of expenses. A simple ROI analysis tallies all of the costs and benefits and compares them. The ROI analysis can be based on the *payback period* or the *net present value* method.

- *Payback period.*   This is the time frame over which the cost incurred would be paid back by the additional cash flow generated by the project. The major drawback is that it does not consider the time value of money.
- *Net present value.*   This takes into account the time value of money, but the main point of controversy is the discount rate. It is normally taken as the weighted average cost of capital. However, if the project is big, the specific cost of allocating money for the project must be considered.

When performing the cost benefit analysis, it is important to be a little conservative in estimating the costs (take the higher estimate) and benefits (take the lower estimate).

PROBLEM DEFINITION

The problem definition clearly specifies the problem that should be attacked by the project team. The problem is defined without delving into the causes, resisting all tendencies to solve the problem before it is defined. It has been noted that teams often start working on a problem only to find out later, after wasting significant resources, that they should have been working on something else. The identification and validation of the problem's causes are steps reserved for later phases of the DMAIC process. The problems could be defined in terms of *what, where, when,* and *how much:* "ABC Company spent 10 percent [how much] of annual sales in fulfilling warranty obligations to its customers in fiscal year 2003 [when] due to marginal modules on the newly released product XXX (in the Latin American Market)."

PROJECT SCOPE

The project scope communicates the boundaries of the project. The scope includes the area, time, duration, seriousness, aspects of the product or service, process boundaries, locations, countries, and product lines. It tells what is within the purview of the project and what is not. To specify the scope of a project, the project must be clearly defined. Sometimes, to clearly understand the scope of the project, the problem definition is changed. Sometimes the project scope crosses multiple departments of divisions, which makes it unmanageable. In such cases, the project scope is right-sized. The management must have a clear definition to enlist the right expertise, and the right scope. Sometimes a project may be split into smaller projects due to its scope. The project scope considers the following points:

◗ Effect on the customer or business   Criticality of the issue
◗ Cost to solve or reduce the adverse conditions
◗ Time commitment
◗ Interdepartmental dependence

The project scope could include project types, business locations, specific warranty failures, limited customer base, and failures observed over a specified time period that accounted for 60 percent of the total field failure cost.

GOAL STATEMENT

The goal is what the project is expected to deliver at the end of the project period. It is the end of the journey as envisaged at the beginning of the journey. This is the motivation for the team to succeed. Six Sigma thrives on dramatic improvements rather than incremental improvements. Therefore, the goals must challenge the project team members' conventional thinking and require them to innovate by not accepting the status quo. Typically, Six Sigma projects target quality improvement of 50–75 percent over the current state of affairs. Motorola had a goal of 68 percent reduction in defects and a 50 percent reduction in the cycle time over 2 years. During the Define phase, the team must gain consensus for common goals that enlist all members of the project team. Examples of the goal may be as follows:

- Reduce the manufacturing lead time from 7 days to 3 days.
- Increase the yield from 60 percent to 99 percent.
- Increase customer satisfaction.

Roles of the Team Members

A team is a group of people working toward a common goal. It works on the basic principle team—that is, *together everyone achieves more* (TEAM). Let's analyze each word here:

Together	Each member utilizes other members' strengths. Team members learn about each other's styles, interests, strengths, and weaknesses. Team members work together, caring for each other's success. They know what others are up to. One is for all, and all are for one. Celebrations for accomplishments and learning lessons from failures are everyone's job.
Everyone	All members are important and have a role to perform in order to move toward a declared goal. Unlike a committee, where a few people work, in the team environment each member encourages the others to perform.
Achieves	Team members are together to achieve on a personal and team level. At the personal level, each member completes assigned action items; at the team level, each one is trying to achieve the common goal. There would be a shared sense of accomplishment of the goal, and at present this is a strong motivational force for all the members.
More	This signifies the basic principle of the synergy—that is, the net result of working together would be more than the sum of the results of individuals working independently.

A team normally works to accomplish a task that one member cannot possibly accomplish because of any of the following reasons:

- The task may be big enough to be handled by one person.
- The task may be very complex, requiring cross-functional skills that no single individual has.

- The task may require the buy-in of all the functions that are affected by the process under review.

In any Six Sigma project, there can be team members at any level, from executives to line personnel, and in between. The roles of various team members in a Six Sigma project are shown in Fig. 4.25.

PROJECT TIMELINE

No activity can produce any economic result if it is not finished when it is needed. With many projects, the team members are not able to prioritize, and the project ends up being canned before it is completed because it is no longer needed. For example, a company was having a production problem for many years. When the problem was finally solved, someone already had made the decision to replace the production line with an

PARTICIPANT	DESCRIPTION OF ROLE
Executive leadership	Provide the driving force behind the adoption of the Six Sigma philosophy and inspire the organization from the beginning of the project.
Deployment Champions	Provide leadership and commitment at the business level. They work to implement Six Sigma throughout their respective businesses.
Project Champion	Oversees supports, provides resources, and funds the Six Sigma project across the organization.
Master Black Belt	Guides and mentors Black Belts in working on the projects. The ultimate expert on Six Sigma at the organization level.
Black Belts	Identify and lead the Six Sigma projects. They have a thorough knowledge of all the tools and techniques of Six Sigma.
Process owners	Provide the necessary cooperation to the Six Sigma projects, ensuring change without disruption perceived by the customers.
Green Belts	Support the Black Belts in the Six Sigma projects. They have the working knowledge of all the tools and techniques of Six Sigma.

FIGURE 4.25.   Team members and their roles.

alternative process. All the work put into solving the problem was wasted.

Six Sigma projects are no exception to this rule. Normally, Six Sigma projects last about 3 to 5 months. If a problem is well defined, the scope is clearly specified, and appropriate resources are available, a project duration of 90 days would be ideal. However, there is no set plan for the timeline. The objective is to solve the problem in an accelerated manner due to the allocated resources. Six Sigma projects must plan for innovative activities. Corporate leadership, once committed to Six Sigma, can hardly wait to see results. The sense of urgency should become a way of life.

In establishing project milestones, one must consider the following:

- *Credibility is critical.* People tend to lose interest if the project drags too long, and this may affect the project outcome.
- *Success talks.* Successful projects generate positive energy, while the failures, even though they are learning opportunities, publicize shortcomings and sap energy.
- *Team confidence.* Personal accomplishments improve self-image and renew commitment to the success of the team.

All action items must have a clearly stated completion date to aim for. Otherwise, one person's ASAP is different from another's ASAP. Project progress must be reviewed at planned intervals. Team members must avoid arbitrarily changing the completion date. There must be deadlines.

RESOURCE REQUIREMENTS

A project can only be successful when enough resources are allocated to execute the plan. Personnel resources must be qualified and must know the project priority clearly. Any deviation from the priority would result in project delays. Resources include materials, people, and tools. Proactively, one can anticipate problems or bottlenecks and prepare resources accordingly. Typical resources required for the Six Sigma projects include the following:

- Office equipment, such as computers, printers, and lab equipment
- Office space
- Lab space
- Communication equipment, such as phone, fax, and Internet service
- Other utilities and office supplies
- Project-specific items

To capture the Define phase, or identify problems, one must listen to the stakeholders, including customers. Identify their requirements and what they consider critical. The Kano model is an excellent tool for analyzing customer requirements, and it facilitates a stronger customer–supplier relationship. Companies have many opportunities for improvement. Pareto charts can be used to prioritize various opportunities based on their internal or external effects. A process map is a great tool to identify internal opportunities for improvement and non-value-added activities. SIPOC tables help to identify various players in the specified theater of operations that could be drivers or restrainers. The project charter is the final deliverable of the Define phase that leads the project team into the phase of understanding the problem better through various performance measures.

UNDERSTANDING THE SCOPE OF THE PROBLEM— MEASURE PHASE

ABCS OF STATISTICS

Pick up a magazine, review a report, or listen to a news program on TV, and you will see the following kinds of statements:

- Food and drink accounts for 50 percent of spending in developing countries; 19 percent in developed.
- Hong Kong, for instance, had 28 times as many personal bankruptcies in 2002 as in 1998.
- In the year ending January 2004, America's industrial production rose by 2.4 percent, the fastest 12-month figure since 2000.

The numbers and percentages that you see in the preceding statements (50 percent, 19 percent, 28 times, 2.4 percent) are called *statistics*. This term refers to numerical facts. Statistics is the art and science of collecting, analyzing, presenting, and interpreting data to understand the business and economic environment in order to make informed decisions.

This chapter introduces some of the fundamental concepts of statistics and preliminary data analysis, which enable analysts to do the following tasks:

- Define data and data sets.
- Understand elements, variables, and observations.
- Discuss scales of measurement.
- Understand data obtained from existing sources or through survey and experimental studies designed to obtain new data.
- Differentiate between qualitative and quantitative data, cross-sectional and time-series data, and discrete and continuous data.
- Summarize data using tabular and graphical methods, including histograms.
- Conduct exploratory data analyses using scatter diagrams.
- Summarize continuous data using numerical methods, including measures of location, variability, and association between two variables.

Data and Data Sets

Data are the facts and numbers that have been collected, analyzed, and summarized for presentation and interpretation to clarify a situation and enable decision making. The data collected in a study are referred to as the *data set* for the study.

Figure 5.1 shows a data set containing comparative information on five low-cost airlines.

Elements, Variables, and Observations

Elements are the objects on which data are collected. For the data set in Fig. 5.1, the five airlines are the five elements of the data set on which data has been collected. A *variable* is a measure of interest against which data is collected. The data set in Fig. 5.1 has the following five variables: *founded, fleet, hubs, unions,* and *assigned seats.* The data collected for each element against each measure are called an *observation.* Thus, with 5

AIRLINE	FOUNDED	FLEET	HUBS	UNIONS	ASSIGNED SEATS
Southwest	1971	Boeing 737	No	Yes	No
Air Tran	1993	Varied	Yes	Yes	Yes
Jet Blue	2000	Airbus A320	No	No	Yes
Song	2003	Boeing 757	No	Pilots only	Yes
Ted	2004	Airbus A320	Yes	Yes	Yes

FIGURE 5.1. Comparison of low-cost airlines. (Data from *Fortune*, March 8, 2004)

elements and 5 variables, Fig. 5.1 has a total of 25 observations (elements × variables).

SCALES OF MEASUREMENT

Many professionals relate the term *data* to numerical data. It is not always so. Data often come in the form of words, letter, or numbers that are mere representations of words or letters. Different types of data can be categorized in one of four scales of measurement: nominal, ordinal, interval, and ratio. The scale of measurement determines the amount of information contained in the data and gives an indication about the appropriate data analysis and presentation. Figure 5.2 provides a summary of the scales of measurement.

Nominal Scale.   The scale of data is *nominal* when the data collected are names or labels to measure the variable for an element. The observations under variable *fleet* in Fig. 5.1 are an example of nominal data. Nominal data could be a nonnumeric label, or a number code to represent a nonnumeric label. For example, in Fig. 5.1, 1 can replace Boeing 737, 2 can replace Airbus A320, and so on.

Ordinal Scale.   The scale of data is *ordinal* when the data exhibit the properties of nominal data, and the rank of the data is meaningful. For example, responses to an airline passenger survey could be ranked on a scale of 1 to 5 in increasing order of passenger satisfaction. Ordinal data can be numeric or nonnumeric.

Scale of Measurement	Description	Example	Type of Data
Nominal	Data collected are names or labels.	Names of countries that are members of the European Union	Qualitative
Ordinal	Data exhibit the properties of nominal data, and the rank of the data is meaningful.	Survey results	Qualitative
Interval	Data have the properties of ordinal data, and the interval between observations is meaningful.	GMAT score	Quantitative
Ratio	Data have the properties of interval data, and the ratio of observations is meaningful.	Height of a group of students	Quantitative

FIGURE 5.2. Summary of scales of measurement.

Interval Scale. The scale of data is *interval* when the data have the properties of ordinal data, and the interval between observations is meaningful. Interval data are always numeric. For example, GMAT scores of 600, 580, 640, and 700 for four students can be ranked from low to high, and the differences between the scores are meaningful.

Ratio Scale. The scale of data is *ratio* when the data have the properties of interval data, and the ratio of observations is meaningful. For example, consider the following data on the weight of four students from the graduating class of the local high school: 120, 130, 140, and 240 pounds. Here, the difference (10) between the weight values 130 and 120 is meaningful, and so is the ratio (2) between 240 and 120. Other common examples of data that use the ratio scale of measurement are height, time, and money. A requirement of this scale is that it must contain a zero value, indicating that the variable has no value at the zero point.

QUALITATIVE DATA

Data that use either the nominal or ordinal scale of measurement are called *qualitative* data. These are the labels and names that identify a variable for each element. Qualitative data can be numeric or nonnumeric. They present a limited opportunity for statistical analysis. Qualitative data can be chiefly summarized as the count or proportion of observations in each category. Even with a numeric code, arithmetic operations such as addition, subtraction, multiplication, and division on qualitative data do not provide meaningful results. One typical example is survey results. On a survey form, six airline passengers responded as follows to a question about their level of satisfaction with the airline's customer service (from 1 to 5 in increasing order of passenger satisfaction): 2, 3, 3, 2, 3, 5. Some of you may summarize this data as an average rating of 3, which is incorrect. This data could be summarized by count (*two passengers gave a rating of 2*) or by proportion (33.33 *percent of passengers gave a rating of 2*).

QUANTITATIVE DATA

The data that use either the interval or ratio scale of measurement are called *quantitative* data. These are numeric values that indicate how much or how many. Quantitative data present many more possibilities for data analysis and presentation. Arithmetic operations such as addition and subtraction provide meaningful results.

DISCRETE AND CONTINUOUS QUANTITATIVE DATA

Quantitative data can be discrete or continuous. Discrete data have a separation between them, while continuous data are literally continuous in nature. As a rule of thumb, if you can describe data by asking how many, they must be discrete data, and if you can describe them by asking how much, they must be continuous data. Good examples of discrete and continuous data are the number of employees in ABC Company's Chicago office and the volume of wine in a bottle, respectively. While you would use "how many" to ask about the former (How many employees work from your Chicago office?), you would use "how much" to ask about the later (How much wine is left in the bottle?).

CROSS-SECTIONAL DATA

Data collected at approximately one point in time are called *cross-sectional* data. The data in Fig. 5.1 are cross-sectional because they describe the five variables for the five airlines as of February–March 2004.

TIME-SERIES DATA

Data collected over several days, weeks, months, or years are called *time-series* data. Time-series data help in monitoring performance, forecasting, and identifying patterns. The time-series graph in Fig. 5.3 shows the seasonality in sales data for ABC Company.

Many reports published by companies have both time-series and cross-sectional analysis. You might have seen company pre-

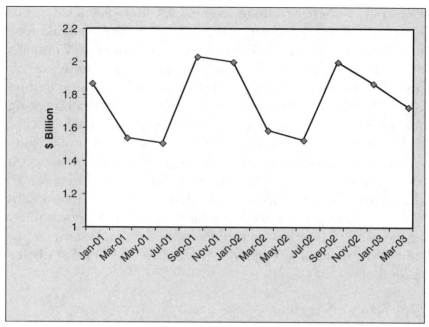

FIGURE 5.3. Quarterly sales for ABC Company.

sentations with graphs showing annual sales trends (time series) and pie charts presenting proportional shares of annual sales by region for the most recent year (cross-sectional).

DATA SOURCES

Collecting good data is a very important task, which is always underestimated. Without good data, your analysis is subjective. Obtaining good data requires careful planning about what is required, why it is required, for how many time periods, and where it can be obtained. Data are available from many existing sources, such as payrolls and sales databases, although there might be situations in which new data have to be obtained through studies and experimentation (for example, conducting a survey in order to obtain data on customer satisfaction with a product or service).

Internal.   All companies maintain a variety of databases about their employees, customers, suppliers, and business operations.

External. There are many companies that specialize in collecting, analyzing, and reporting data. Data are available for a fee. Dun & Bradstreet, Bloomberg, Dow Jones & Company, and Standard & Poor's Market Insight are examples of providers of data and reports. Information from these companies is helpful in market research and industry benchmarking. All industry associations carry industry-specific data on their websites, which is normally available to members, and in some cases to nonmembers as well. Many government agencies provide detailed demographic and trade data. Some examples are the Census Bureau and the U.S. Department of Labor. Figure 5.4 lists examples of data available from some government agencies. The Internet continues to be the single most important and exponentially growing source for data and statistical information.

STATISTICAL STUDIES

When data for the required application are not available from existing sources, the data can be obtained by conducting a statistical study. Statistical studies are classified as either experimental or observational.

Experimental Study. In an experimental study, the variables that affect the variable of interest are identified and controlled

GOVERNMENT AGENCY	EXAMPLE OF DATA AVAILABLE
Census Bureau (www.census.gov)	Population data, income data, and education data
Bureau of Labor Statistics (www.bls.gov)	Consumer spending, earnings rates by profession, unemployment rate, and other international statistics
STAT USA (www.stat-usa.gov)	International market research, trade opportunities, country analysis, current and historical economic and financial releases, and economic data
Federal Reserve Board (www.federalreserve.gov)	Data on discount rates, installment credit, and exchange rates

FIGURE 5.4.   Examples of data available from government agencies.

in order to obtain data on how they influence it. For example, in order to understand the effect of a new process for serving passengers on an airline, the management uses the same group of flight attendants for the old and new processes in the study to monitor customer response to the new service.

Observational Study. In an observational study, no attempt is made to control the causal variables. Various surveys are examples of observational studies.

Statistical studies are expensive means of obtaining data, and require significant time and resources. If conducted with thought and care, they can provide much better results in situations where you are examining the effects of multiple factors on a desired variable, such as revenue. The decision to employ such methods should be carefully evaluated to confirm that the benefits outweigh the costs.

ERRORS IN DATA ACQUISITION

Not using data is sometimes better than using bad data. An error in data acquisition occurs whenever the data obtained differ from the actual values. For example, an interviewee may misunderstand a question and provide an incorrect response, or data-entry personnel may add an extra zero, entering 500 for 50. There are simple processes that you may employ to identify outliers and to ensure that your data are error free. You can quickly review your data to identify obvious errors, such as data that report that an employee is 30 years old and has 32 years of experience.

FREQUENCY DISTRIBUTION

A frequency distribution is the first step toward developing a bar graph. It is a tabular summary of data listing the number (frequency) of items in each of the nonoverlapping classes. Consider the cola preferences of a group of 30 customers at a shopping mall (Fig. 5.5) to demonstrate the construction and interpretation of a frequency distribution for qualitative data. To develop a frequency distribution for these observations, count the number of times each cola appears in the table. This frequency distribution provides a summary of how purchases are distributed across the four colas (Fig. 5.6). Further, you can obtain relative frequency by dividing the frequency of each class by the number of observations.

$$\text{Relative frequency of class} = \frac{\text{frequency of class}}{n}$$

Cola A	Cola D
Cola C	Cola A
Cola B	Cola D
Cola C	Cola A
Cola A	Cola D
Cola B	Cola B
Cola D	Cola B
Cola B	Cola B
Cola C	Cola C
Cola A	Cola C
Cola C	Cola C
Cola A	Cola C
Cola C	Cola C
Cola C	Cola D
Cola C	Cola C

FIGURE 5.5. Cola preferences of 30 customers at a shopping mall.

COLA	FREQUENCY	RELATIVE FREQUENCY
Cola A	6	0.20
Cola C	6	0.20
Cola B	13	0.43
Cola C	5	0.17

FIGURE 5.6. Frequency distribution of qualitative data.

BAR GRAPHS

The frequency distribution for qualitative data can be illustrated with a bar graph. A *bar graph* is simply a graphical representation of frequency distribution or relative frequency distribution. You can use the graph function in Excel or PowerPoint to make bar graphs. Figure 5.7 shows a bar graph for the cola data. The *x*-axis (horizontal) specifies the labels (or categories), and the *y*-axis (vertical) specifies the frequency (or relative frequency). The bars are separated to emphasize the fact that each category is separate (qualitative data).

A pie chart is another graphical device to illustrate frequency and percentage frequency distributions.

QUANTITATIVE DATA

FREQUENCY DISTRIBUTION

Developing a frequency distribution for quantitative data is the same as for qualitative data, except that it is not as simple to

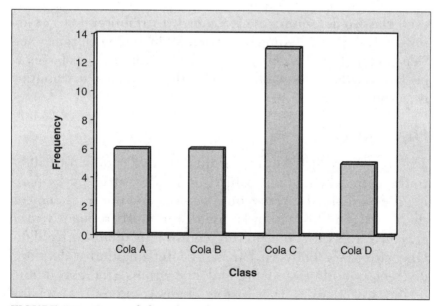

FIGURE 5.7.   Bar graph for cola preference.

identify the classes or categories. The following three steps are helpful in defining classes with quantitative data.

1. *Estimate the number of classes.*   Specify the number of classes required to estimate the frequency distribution. It should be enough to capture the variation in data without having so many classes that some have very few data items. It is recommended to use 5 to 15 classes.

2. *Determine the width of each class.*

Approximate class width = largest data value

− smallest data value/number of classes

3. *Specify class limits.*   Class limits must be chosen so that each data item belongs to only one class. The lower class limit indicates the smallest possible data value that can be assigned to the class, and the upper class limit indicates the largest possible data value that can be assigned to the class.

Once the number of classes, class width, and class limits are available, count the number of data values belonging to each class to determine the frequency distribution. Let's examine the call handling time data from 50 incoming calls at a service center (Fig. 5.8). If we consider the number of classes to be 10, and the class width to be 8, the frequency distribution is as shown in Fig. 5.9.

HISTOGRAMS

The frequency distribution for quantitative data can be illustrated with a histogram. A histogram is constructed by placing the variable of interest on the x-axis and the frequency on the y-axis. The histogram is one of the most frequently used graphical tools for interpreting continuous quantitative data. One difference between bar charts (for qualitative data and discreet quantitative data) and histograms (for continuous quantitative data) is that histograms have no natural separation between the rectangles of adjacent classes. This can be

29	21	1
5	22	2
22	14	10
4	7	2
10	47	33
22	61	5
52	72	4
20	22	53
5	10	15
26	23	37
7	8	8
80	2	4
20	51	20
5	5	65
58	32	4
16	13	10
37	3	

FIGURE 5.8. Call handling time in seconds for 50 incoming calls at a service center.

CLASS	FREQUENCY	RELATIVE FREQUENCY
0–8	18	0.36
9–16	8	0.16
17–24	9	0.18
25–32	3	0.06
33–40	3	0.06
41–48	1	0.02
49–56	3	0.06
57–64	2	0.04
65–72	2	0.04
73–80	1	0.02

FIGURE 5.9. Frequency distribution of quantitative data.

achieved in Excel by double-clicking on the histogram bars, selecting Options, and selecting 0 for Gap Width. Figure 5.10 shows the histogram for the call handling time data. How would you interpret this histogram? A quick glance shows that the data is right skewed and has a very high variation, although a majority (70 percent) of the calls are within the first three class intervals.

EXPLORATORY DATA ANALYSIS

So far this chapter has shown how to summarize and represent one variable data using frequency distribution tables, bar graphs, and histograms. The exploratory data analysis uses simple tools to determine the relationship between two variables. This section discusses scatter diagrams, a commonly used tool for such analysis. The relationship between advertising spending and company sales is an example of a two variable analyses.

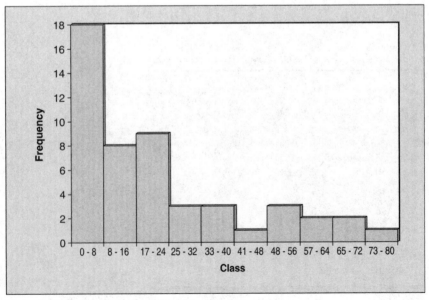

FIGURE 5.10.   Histogram of call handing time data.

SCATTER DIAGRAMS

A *scatter diagram* is a graphical presentation of the relationship between two quantitative variables. It is developed by plotting the dependent variable (or the variable of interest) on the x-axis and the independent variable (the variable that affects the dependent variable) on the y-axis of a two-dimensional graph. Figure 5.11 lists the weekend sales and the preceding week's advertising expense data for a neighborhood grocery store. The objective is to evaluate the effect of advertising on sales. Figure 5.12 shows the scatter diagram based on the grocery store sales data. If you were the store's manager, what conclusions could you draw from looking at the graph? The upwardly clustered observations indicate that the

WEEKLY ADVERTISING EXPENSE	WEEKEND SALES
$10,000	$100,000
10,500	100,100
11,000	100,500
11,500	99,500
13,000	105,000
15,000	112,000
20,000	124,000
25,000	128,000
30,000	130,000
29,000	129,700
28,000	129,400
27,000	130,000
27,500	130,600
28,000	131,200
29,000	131,800
30,000	131,200

FIGURE 5.11. Advertising expenses and sales returns for local grocery store.

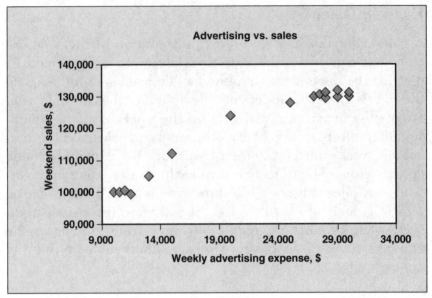

FIGURE 5.12.   Advertising expenses and sales returns.

sales growth is positively related to advertising expense. You can also see that the graph plateaus at about $25000 weekly advertising expense. Although scatter diagrams are a great tool for quickly seeing the trend between two quantitative variables, they may not be able to explain all the other factors involved that might have affected sales. For a more concrete model, advanced statistical analysis such as regression is required. Figure 5.13 shows scatter diagrams for positive, negative, and no relationship between two variables. Scatter diagrams can be generated with the graph wizard in Excel.

SUMMARIZING DATA: NUMERICAL METHODS

If you are presented a quantitative data set, what is your first inclination? The first thing I want to do is summarize it so that I make some sense out of it. Data can be summarized by gauging their location on a scale and by quantifying their variation. Here are some of the measures of location and variation. Also included is the measure of association between two variables (recall the discussion on scatter diagrams).

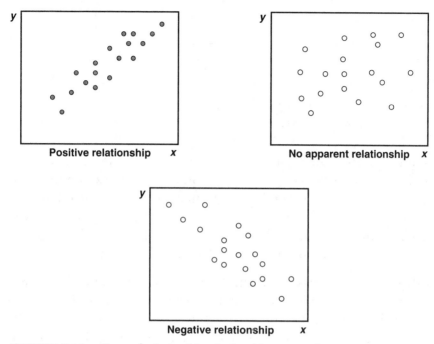

FIGURE 5.13.   Types of relationships depicted by scatter diagrams.

Mean.   The *mean* or average is perhaps the most important measure of location for a variable. The mean provides a measure of central location for the data. There are different notations for data from a sample (just a portion of the data), and data from a population (all the data). The data from a sample are denoted by $\bar{x}$, and the data from a population are denoted by the Greek letter μ.

Sample Mean

$$\bar{x} = \frac{\Sigma X_i}{n} \qquad \text{for } i = 1 \text{ to } n$$

where $\Sigma X_i = X_1 + X_2 + X_3 + X_4 + \cdots + X_n$
n = sample size

The Greek symbol Σ stands for "sum of" (summation).

Population Mean

$$\mu = \frac{\Sigma X_i}{N}$$

where $\Sigma X_i = X_1 + X_2 + X_3 + X_4 + \cdots + X_N$
N = population size

Consider the call handling time data in Fig. 5.8. The average call handling time is 22 seconds. You can calculate the mean or average using the preceding formula. You can also use Excel to do so with a simple command. Enter the following in any of the cells in an Excel worksheet: "=AVERAGE(data cell range for variable x)." For example: "=AVERAGE (A2:A52)."

Median.   The median is another measure of central location for data. While the mean sums the data points and divides the result by the number of observations, the median simply tells you the middle point in your data set. For an odd number of observations, the middle value obtained by arranging the observations in an ascending order is called the *median*. For an even number of observations, the median is the average of the two middle values. Although the mean is the more commonly used measure of central location, in some situations, such as whenever data have extreme values, the median is the preferred measure of location. For example, note that the government gives out the median annual family income.

Median (using Excel): "=MEDIAN(data cell range for variable x)"
Median (for data in Fig. 5.8) = 16

Mode.   The mode is the value that occurs with greatest frequency. For example, the value that occurs most frequently in the call handling data set is 5. You may encounter situations with two or more different values with the greatest frequency. In these instances, more than one mode exists. If the data have exactly two modes, we say that the data are bimodal.

Mode (using Excel): "= MODE(data cell range for variable x)"

Range.   The range is simplest measure of variability. It is the difference between the largest and the smallest values in the dataset. It has limited application because of its sensitivity to just two of the observations.

Range (using Excel): "= max(data cell range for variable x)"

for largest value

"= min(data cell range for variable x)"   for smallest value

Range = largest value − smallest value

Range (for data in Fig. 5.8) = 79

Variance.   The variance is a measure of variability that utilizes all the data. The variation is the difference between the value of each observation x_i and the mean (also called the deviation about the mean). Variance is the average of the squared deviations.
Population variance is denoted by Greek symbol σ^2

$$\sigma^2 = \frac{\Sigma(x_i - \mu)^2}{N}$$

Sample variance is denoted by s^2

$$s^2 = \frac{\Sigma(x_i - x)^2}{n - 1}$$

Note: If the sample mean is divided by $n - 1$, and not n, the resulting sample variance provides an unbiased estimate of the population variance.

Variance (using Excel): "=VAR(data cell range for variable x)"
Variance (for data in Fig. 5.8) = 429

Standard Deviation.   The standard deviation is the square root of the variance.

$$\text{Sample standard deviation} = s = \sqrt{s^2}$$
$$\text{Population standard deviation} = \sigma = \sqrt{\sigma^2}$$

Variance (using Excel): "=STDEV(data cell range for variable x)"

Variance (for data in Fig. 5.8) = 21

Correlation Coefficient.   Consider the discussion of scatter diagrams. Now that you have an understanding of the association between sales and advertising dollars in the example discussed earlier, you may want to quantify this relationship. The correlation coefficient provides the magnitude of linear relationship between two variables. The value of the correlation coefficient lies between 0 and ±1. A correlation coefficient of +1 indicates a perfect positive linear relationship, and a value of −1 indicates a perfect negative linear relationship. A value of 0 indicates that there is no linear relationship between the two variables.

Correlation Coefficient for Sample Data

$$r_{xy} = \frac{s_{xy}}{s_x s_y}$$

where s_x and s_y are standard deviations for variables x and y, and s_{xy} is the sample covariance.

Correlation coefficient (using Excel): "=CORREL(data cell range for variable x, data cell range for variable y)"

Correlation coefficient (for data in Fig. 5.11) = 0.98

STATISTICAL THINKING

NORMAL DISTRIBUTION

If you are reading this book, chances are you have already heard about normal distribution. Before we define normal distribution, consider how it helps us. Say that you are a Six Sigma consultant with a bank, and you have been assigned the job of reducing the customer service cycle time at the teller counter, that is, from the time the customer approaches the counter until the customer leaves the counter after receiving service. The first thing you do is measure the current cycle time. You spend a few days watching the process closely and measure the cycle time using a stopwatch. You record cycle times for 50 customers over a period of 1 week. You randomly pick a different customer service agent every shift, because you want to represent normal service experience by the customer across all shifts. Now that you have the data, you can estimate the process performance. You should remember how to calculate the process mean from the basic statistics discussion. However, the process mean does not tell the entire story. It does not tell you about the process inconsistencies or variation. Two processes with the same mean can have entirely different variations. To estimate variation, you can calculate the process standard deviation. To illustrate mean and variation, you can construct a histogram. While the histogram shows you the spread of the distribution, it does not allow you to make process performance estimates. This is where the normal distribution comes in handy. The normal distribution allows you to estimate the proportion of process values that would fall within a certain range.

So, what is normal distribution? It is a bell-shaped distribution of values in which the majority of the items are in the center and some are on the tail ends. The frequency of occurrence of values decreases symmetrically on either side of the mean as you go further away from it. The normal distribution construction is based on the process mean and the process standard deviation. The distribution looks like a bell-shaped curve (see Fig. 5.14). The area under the curve helps to define what proportion of values will fall within a certain range. For example, empirically we know that 68.26 percent of all values fall within ±1 standard deviation (SD), and 95.46 percent of all values fall within ±2 SD. For the bank customer service cycle time example, if the process mean is 6 minutes and the process standard deviation is 1 minute, and assuming that the process follows the normal dis-

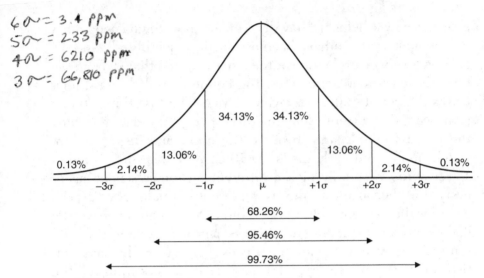

6σ = 3.4 ppm
5σ = 233 ppm
4σ = 6210 ppm
3σ = 66,810 ppm

34.13%   34.13%

13.06%   13.06%

0.13%   0.13%
      2.14%   2.14%

  −3σ   −2σ   −1σ   μ   +1σ   +2σ   +3σ

68.26%

95.46%

99.73%

FIGURE 5.14. Normal distribution probability curve.

tribution (which it would unless there are special cause variations), we can expect that 68.26 percent of all process cycle time
values will be between 5 minutes and 7 minutes. As we just did,
if you know the process is normally distributed, and if you have
the process mean and process standard deviation, you can estimate process performance. If you would like to, any statistical
software can help you construct a normal distribution.

How do we know that the process under consideration follows a normal distribution? Most of the processes do. Height
data for a random sample of men or women in your company
would follow a normal distribution. A simple way to investigate
normal distribution is to construct a bar graph or a histogram
with the process data. If the histogram follows the properties of
a normal distribution (centered on the mean, symmetrically
distributed), the data are most likely normally distributed.
There are statistical tests to determine normality, which are
beyond the scope of this text. You may refer to any business statistics book to find out more about normal distributions.

EXAMPLE APPLICATION

Going back to the bank customer service cycle time example,
Fig. 5.15a illustrates the distribution of this process with a

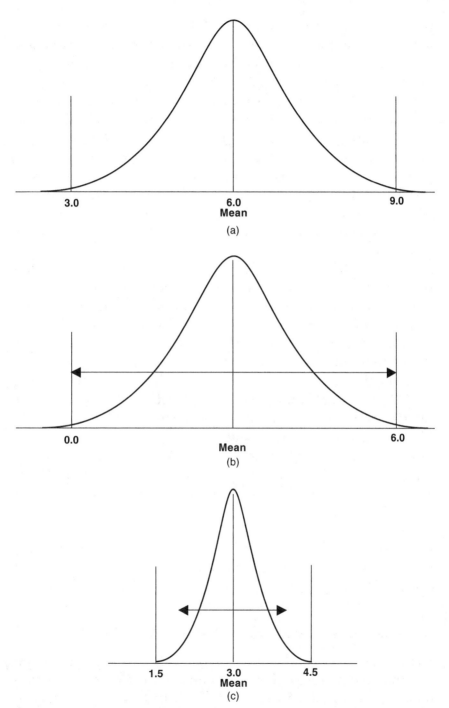

FIGURE 5.15. Bank customer service cycle time: (*a*) current performance (SD = 1), (*b*) reducing mean by 50 percent (SD = 1), and (*c*) reducing variation (SD = 0.5).

process mean of 6 minutes and a standard deviation of 1 minute. As per the properties of a normal distribution, we know that 99.73 percent of all process values will be within 3 to 9 minutes (±3 SD on both sides of the mean). Is this acceptable? Say the bank management wants to reduce the cycle time to 3 minutes in order to increase customer service capacity and reduce waiting time. What would you like your process performance to look like? Many may simply suggest moving the process mean to 3 minutes. (Of course, there are various steps involved in figuring out the ways to reduce the process mean, which is discussed in other sections.) Suppose you are able to achieve a reduced process mean of 3 minutes. The new process looks something like the curve in Fig. 5.15*b*. It is centered on 3 minutes. The process variation (standard deviation) is 1, and so 99.73 percent of all process values now fall between 0 and 6 minutes, and 50 percent of all values fall beyond 3 minutes. You will probably go back to management and present this new situation. Assume that they suggest that any cycle time greater than 4.5 minutes is too high. What options are left to you? You could further move your average to less than 3 minutes, so that 99.73 percent of values would fall within 4.5 minutes, or reduce variation so that the same result could be achieved. Moving the average further might be difficult, but reducing the process variation might be possible. Figure 5.15*c* shows that by reducing the variation (standard deviation) by 50 percent (from 1 to 0.5), you could achieve a majority of process values within 1.5 to 4.5 minutes.

In short, a normal distribution helps you in making decisions about the process based on the sample data. The prerequisite, however, is that you have to determine whether normal distribution applies to your process.

RANDOM VERSUS ASSIGNABLE VARIATION

Variation in data can be of two types, one that is common and one that occurs rarely. The common variation that occurs randomly constitutes the normal or bell-shaped distribution. Statistical thinking requires a clear understanding of the difference between random and assignable variation. Random variation occurs due to uncontrolled variables that exist in the environment—for example, fluctuation in temperature and relative humidity in a manufacturing facility, or normal variation in driving time commuting to work or back to home. Assignable variation occurs due to a specific cause. For example, getting home an hour or so late may be caused by a traffic jam due to an accident, or by a police car parked on the roadside.

When the chance of an event occurring is small, and the event occurs, it is considered an assignable cause. This means something specific has occurred. On the other hand, when some event occurs routinely, normally, or regularly, it becomes a common event, and it is no longer controllable. This explanation initially seems counterintuitive, because we think that when an event occurs rarely, it only happens by chance (implying that there is only a small chance of it ever happening). In statistical thinking, *everything* happens by chance; that is, there is a probability that it will happen. It is the *degree* of chance that separates random events from assignable events.

Figure 5.16 shows the difference between random and assignable variation. The less likely aspect of variation is considered assignable, while the more likely aspect of variation is considered random. The causes that contribute to random variation are difficult to identify, difficult to control, and challenging to change. The causes that contribute to assignable variation are easier to identify (assign) and easier to control or change.

In applying various statistical techniques such as control charts, design of experiments, C_p, or C_{pk}, or in problem solving, managing processes, or determining your responses, you can make use of knowledge of random or assignable variation or causes. If the variables are random and cannot be controlled, your response would be different. Similarly, if you know that the variation is assignable, with known causes, your action to change the state of the variable would be different from your response without such knowledge.

In controlling a process, the objective is to maintain the process in statistical controls (i.e., without any assignable variables). When improving a process using a test hypothesis,

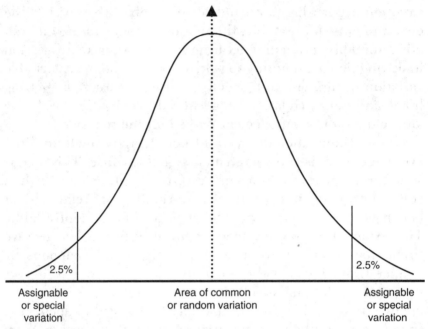

FIGURE 5.16.   Random versus assignable variation.

analysis of variance (ANOVA), or design of experiments, the objective is to determine whether the effects of the process change are significant enough that they have a small probability of occurrence (unless planned or assignable), or whether the probability of their occurrence is common, and therefore uncontrolled (not as planned or random).

DIGGING FOR GOLD

STARTING AT HOME

Poor quality costs money. Consciously or unconsciously, we are familiar with this concept. Imagine a weekend dinner when you have invited some special guests. Your guests love spicy seafood, so you decide to try your hand at Tom Som Pla, a Thai dish made with fish steamed with lemongrass, from an Asian recipe book you recently bought. All goes well, except that you oversteamed the fish. It turns gooey, unfit to be served to your guests. Time is short before the guests arrive. What do you do? You probably call the neighborhood Asian restaurant for a quick delivery. Did you ever estimate the total cost to you because of this kind of mishap? Maybe not. After all, you don't cook exotic dish often, and neither do you invite guests over for dinner every day. However, just for fun, let's examine what could be your total cost in a situation like this:

- The cost of the ingredients.
- The cost of any special equipment required for this dish.
- You and your spouse work on the dish for 4 hours each, including the time spent browsing the recipe book, shopping for ingredients, and, of course, cooking. That is a total of 8 person-hours.
- The cost of ordering the substitute from the restaurant, including the phone call, delivery charge, tax, and tip.
- The cost of the extra stress to both you and your spouse because of the mishap, and the resulting effect on your readiness to receive your guests.
- The cost of lost opportunity. Had you planned to order from the restaurant to begin with, you could have utilized those 8 hours and the cost of the ingredients on something more productive.

If you add all of this up, it will be a substantial sum. Imagine if you were running a company and this was a more frequent occurrence. Believe it or not, many companies, large and small, do end up producing poor-quality products. By some estimates most companies spend 20 to 30 percent of their revenues *repairing or reworking* products and processes. After the initial gasp, the natural reaction of most people to these numbers is disbelief. As we try to understand what constitutes this 20 to 30 percent, let's analyze the costs from the failed Tom Som Pla experiment.

DESCRIPTION	TYPE OF COST
Cost of ingredients	Internal failure
Cost of equipment	Internal failure
Time to buy ingredients and cook the dish	Internal failure
Placing order at the restaurant	Rework
Stress (the stress may lead to unseemly outcomes, such as abrupt behavior on your part, thus displeasing your guests)	External failure
Using resources for more productive work (like doing the yard work you had been planning for so long)	Opportunity cost

WHAT HAPPENS AT WORK?

In almost every organization, there are many situations in which a job is done incorrectly, leading to quality defects, delayed shipments, and rework. When discovered at a later stage in the process, such errors have the reverse of an exponential effect in terms of finding where the mistake occurred. This leads to emergency expenses such as getting everyone together for meetings to address the issue, dealing with dissatisfied clients, paying overtime, making overnight shipments, having to fly people to different locations, and so on (see Fig. 5.17a for cost of quality components). Besides the extra expense, there is a huge opportunity cost. Everyone's schedule is blown, and other important assignments are pushed to a later time. While most companies track the rework and warranty costs associated with poor quality, very few, if any, measure the person-hours, travel costs, dissatisfied customers, dissatisfied employees (late hours and extra stress affect their family lives), productive work that could have been accomplished in the same time, and other such opportunity costs (Fig. 5.17b). Besides, the cost to the customer could be enormous. Consider an auto parts supplier that cannot deliver quality bearings on time for a new design being launched by a major auto company, thus delaying the launch schedule.

Many companies are now taking the approach of calculating costs based on a particular activity. A major technology and

INTERNAL FAILURE	APPRAISAL	CURRENTLY MEASURED
✓ Revise, rework, review	✓ Inspection	✓ Warranty costs
✓ Firefights	✓ Testing	✓ Compliance incidents
✓ Opportunity cost	✓ Quality audit	✓ System downtime
	✓ Initial cost and maintenance of test equipment	✓ Production rejects
EXTERNAL FAILURE	PREVENTION	NOT MEASURED
✓ Warranty costs	✓ Quality planning	✓ Opportunity cost
✓ Complaint adjustments	✓ Process planning	✓ Revise, rework, review
✓ Compliance issues	✓ Process control	✓ Quality checks
✓ Cost to customer	✓ Training	✓ Increased maintenance
✓ Opportunity cost		✓ Lost accounts
		✓ Customer dissatisfaction
		✓ Development errors
(a)		(b)

FIGURE 5.17. Cost of quality: (*a*) in components, and (*b*) measured versus not measured.

management consulting firm traces all costs associated with its sales activity by each account, including market research, proposal development, and travel costs, in order to ensure that the sales effort is worth the resources in the short or long term. A sales manager would think twice before flying seven consultants from New York to the West Coast on a sales presentation for a $100,000 contract, unless this contract was a precursor to larger contracts in the future. Similarly, companies can deploy methods to track resources invested in various activities, projects, and processes.

ACTIVITY-BASED COST ACCOUNTING

Activity-based cost (ABC) accounting is one method for tracking resources. ABC accounting tries to identify the real cost associated with serving each entity (customer, project, or process). Both the variable costs and the overhead costs are decomposed and tagged back to the entity. Companies that fail to measure their real costs correctly are not measuring their profits correctly. Identifying the true costs associated with a sale, a project, or a process emergency over a period of time will also help to highlight the problems associated with rework and firefighting. (*Please refer to any corporate finance book for more details on ABC.*)

EFFECT OF POOR QUALITY

In addition to cost overruns and reduced profitability, poor quality also affects employee morale. We touched upon the stress that comes from working late hours trying to mitigate a situation that might blow up. Quality failures can lead to finger pointing among employees. The high frequency of mishaps may encourage employees to avoid responsibility for fear of the backlash. Lots of firefights and customer complaints will also erode their esteem for the work they do. Overall, it may result in a very negative work atmosphere.

Poor quality can have even more serious effects on your revenue and brand value. In the short run, you may avoid repercussions from poor quality through promises, warranty payments, and other such adjustments. However, this would affect your revenues and profitability in the long run. Most companies have 5 to 10 percent of customers who are dissatisfied with its products, sales, or service. Not only do they lose the existing customer, they can lose many more potential customers by word-of-mouth communication. Studies have shown that although only 1 out of every 25 dissatisfied customers complains to the company, all 25 will mention their dissatisfaction to 7 to 10 other potential customers.

WHAT CAN BE DONE?

The rest of this book discusses Six Sigma and its role in improving business performance. The underlying principle of the Six Sigma approach is *thinking differently* (Fig. 5.18). When the discussion turns to quality, managers and employees often

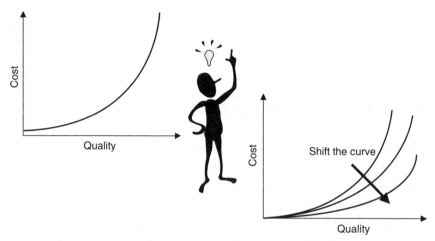

FIGURE 5.18.   Quality cost curves—thinking differently.

become defensive and argue that they have done their best, and that any further resources will only add to the cost. They're right—adding another person just to check the process output will not help! After all, it's only a check; the best it can achieve is to identify the mistake that has already been made. You need to identify the root cause of the problem and eliminate the root cause by thinking out of the box.

Sometimes just analyzing the data and reports that you already have will help. A medical devices company has an extensive quality-control program in which randomly selected pieces are checked before they are moved to the next workstation. This, however, does not reduce quality defects. Not only did the managers continue to employ the same processes, they never actually analyzed the data obtained from the quality checks to identify patterns or irregularities. Had they done so, they would have realized that the process had shifted toward one of the control limits, resulting in up to 50 percent of the pieces falling outside the control limits. We shall discuss this in more detail under process capability.

Most of the errors come from the processes and systems, not from employees. Just focusing on employees will yield very limited results. As W. Edwards Deming has said, "Eighty-five percent of the reasons for failure to meet customer expectations are related to deficiencies in systems and process...rather than the employee. The role of management is to change the process rather than badgering individuals to do better" (Deming, 1988).

MEASURE FOR MEASURE

Measurement is required to determine whether a process is working effectively, or the product is functioning within specifications. Measurement data are collected after verifying a system's performance. Each measurement has some uncertainty associated with it, because nothing is theoretically constant. For example, if one measures the speed of light using two measuring devices, one will get two different answers. Does this mean that the speed of light has changed? No. So, no two systems can measure identically. Instead, the measurement system adds its own variability to the measurement of performance. The measurement system can change in many ways, such as in the measuring equipment, the operator, the parts, or the operating environment. People ask, How accurate is a measurement system? What is its repeatability or reproducibility? The methodology that provides the answers to such questions, that quantifies the variability added by a measurement system, is called *measurement system analysis* (MSA).

Measurements are taken almost everywhere. At home, you measure the heating or cooling temperature, your personal weight or height, your fever temperature, your blood pressure or blood sugar level. Blood pressure measuring systems once tended to be faulty, and when they showed an excessively high reading, the person was immediately sent to emergency care. In the manufacturing environment, equipment has its own process-controlling parameters; parts or products have their dimensions or operating parameters, electrical resistance, current, power, speed, viscosity, diameter, or length. In the service environment, measurement may include service time, or other parameters similar to those in manufacturing. Since parameters can be of the attribute or variable type, so are the measurement systems.

In many businesses, measurements are taken without paying any attention to the quality or consistency of measurements. Sometimes the error introduced by the measurement system is excessive, such that the actual measurement becomes unreliable. In other words, the true value of the measurement may be too far from the measured value. As one plans to achieve Six Sigma–level performance and the process improves, sensitivity to the variation added by a measurement system becomes more predominant. Figure 5.19 shows the various elements of a measurement system. In order to have confidence in the measurements taken, or the data collected, one must perform measurement system analysis.

- *Attribute gages.* These are fixed gages designed to make a go, no-go decision. These gages indicate whether a product is good or bad. Examples include master gages, plug gages, and ring gages.
- *Variable gages.* These instruments measure actual physical dimensions. They provide a measure of how good or bad the product is, relative to the specifications. Examples include line rules, vernier calipers, and micrometers.

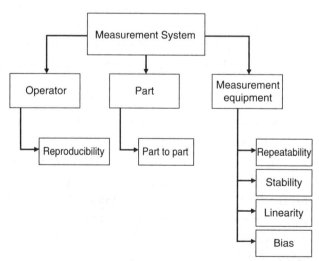

FIGURE 5.19. Measurement system analysis components.

Definition of Terms

Measurement system analysis (MSA).   MSA is an experimental and mathematical method of determining how much the variation within the measurement process contributes to overall process variability. MSA is a critical first step that should precede any data-based decision making. The purpose of MSA is to qualify a measurement system for use by quantifying its accuracy, precision and stability.

Bias.   The difference between the output of the measurement system and the true value.

Sensitivity.   The gage should be sensitive enough to detect differences in measurement as slight as 0.10 of the total tolerance specification or process spread, whichever is smaller. In selecting a measurement system, one must be able to detect the small changes in the measured characteristics faithfully. The resolution of the measuring device must be at least 0.10 of the tolerance for the measured characteristic. For example, if one is interested in measuring people's height where the perceived requirement is to be feet and inches, the resolution should ideally be 0.10 inch. However, for all practical purposes, people like to know in 0.25-inch increments.

Stability.   The ability of a measuring instrument to maintain constant metrological characteristics over a specified time interval. One must be aware of the stability of the process. Consider the example of two newly purchased and calibrated micrometers that are used to measure the dimensions of machined parts. After 3 months, one of them still measures just like a new micrometer, while the other one has changed, and introduces measurement variation.

Linearity.   This is the difference in the bias values through the operating range of the gage. Consider a device that measures the temperature of an oven used in manufacturing, especially during the process setup when the temperature is ramping up. For some diffusion furnaces, the temperature goes as high as 1000°F. The linearity of the measurement system is its ability to maintain reproducibility and repeatability throughout the temperature range, i.e., 0–1000°F.

Calibration. The set of activities to ensure that measuring instruments maintain their metrological characteristics by comparison to a nationally or internationally recognized standard. In calibration the focus is on stability, linearity, bias, and repeatability.

Accuracy. This is an unbiased true value and is normally reported as the difference between the average of a large number of measurements and the true value.

Precision. This is a measure of the consistency of measurements or the standard deviation of the measurements. Figure 5.20 demonstrates the difference between accuracy and precision.

Reproducibility. This is the ability of the gage system or similar gage systems to reproduce measurements. Reproducibility of a single gage is checked by comparing the results taken by different operators. Gage reproducibility affects both accuracy and precision. Reproducibility is the variation in average measurements caused by factors such as operators, gages, or environmental conditions.

Repeatability. This is the ability of a gage system to repeat measurements that are taken under identical conditions. Repeatability is the variation caused by a gage when an oper-

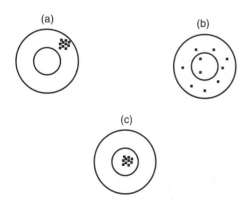

FIGURE 5.20. Precision and accuracy: (*a*) precise, but not accurate; (*b*) accurate, but not precise; and (*c*) precise and accurate.

ator repeatedly measures the same characteristics at the same location on the same part.

MSA FUNDAMENTALS

▫ Determine the number of repeat readings.

▫ Use appraisers who normally perform the measurement.

• There should be a specific, documented measurement procedure that is followed by all appraisers.

• Select the sample parts to represent the entire process spread.

• If applicable, mark the exact measurement location on each part to minimize the impact of within-part variation.

• Ensure that the measurement device has adequate resolution.

• Measurements should be taken in random order. A third party should record the measurements, the appraiser, the trial number, and the number for each part on a datasheet.

R&R ANALYSIS METHODS

Range method.   This is a simple way to quantify the combined repeatability and reproducibility of a measurement system.

Average and range method.   This computes the total measurement system variability, and allows the total measurement system variability to be separated into repeatability, reproducibility, and part variation.

Analysis of variance (ANOVA) method.   This method of performing R&R studies is the most accurate method for quantifying repeatability and reproducibility. It also allows the variability of the interaction between the appraisers and the parts to be determined.

R&R STUDY PROCEDURE

Repeatability and reproducibility of a measuring device is assessed either at a prescribed interval or when its ability to measure effectively is questioned; in other words, when measurements become unbelievably good or unacceptable. The Automotive Industry Action Group (AIAG) has published a

Measurement System Analysis Reference Manual as part of the QS-9000 quality management system. According to the manual, the following procedure is recommended:

Variable Gage

1. Obtain samples of 10 parts that are representative of the process variation. Number them 1 through 10.
2. Select three (or two) appraisers and identify them as A, B, and C.
3. Have appraiser A measure 10 parts in random order. Then let appraisers B and C measure the 10 parts in random order.
4. Repeat Step 3 two more times, randomizing parts each time.
5. Enter the results in a data sheet for further analysis or use a statistical analysis software such as Minitab, Statgraphics, or JMP.

Attribute Gage.   For attribute gages, the R&R study is conducted by selecting 20 parts and two appraisers. Select parts that represent the entire tolerance. One can select some parts outside the specifications as well, to incorporate typical variation in the measurement system. Measure all parts in random order to minimize appraiser bias.

MEASUREMENT SYSTEM DATA ANALYSIS

Once the data are collected, they are analyzed for variation in the measurement system and in its repeatability, reproducibility, and part-to-part variation components. These components of the measurement system are determined as a percentage of the allowed tolerance, or a percentage of the total process variation. If there is a question about the effectiveness or resolution of the measurement system, one must use it with respect to the allowed tolerance. If there is a question about the significance of the measurement system variation, one must compare it with the process variation.

Figures 5.21 through 5.23 show the Gage Reproducibility and Repeatability data collection and analysis worksheets. One can use two or three operators for the Gage R&R study and determine various components of variation, as follows:

PART AND GAGE INFORMATION						
Part Number	03312004-00000		Gage Name	Gram Scale		
Description	TP155 SDW 31.00 M.R.		Gage ID			
Characteristics	Adhesive Coat Weight		Date of Study			
Specifications	12.5 - 17.0 (lbs/ream)		Performed by	Joe and Jane		
STUDY DATA						
Number of appraisers (n)	2					
D4 (3.27 for 2 trials and 2.58 for 3 trials)	3.27					
D3 (0 for up to 7 trials)	0					

FIGURE 5.21. Gage R&R study worksheet—part and gage information.

Appraiser	Trial #	1	2	3	4	5	6	7	8	9	10		Average
A	1	14.74	15.58	14.51	14.5	13.55	15.58	14.27	14.15	15.69	14.26		14.6830
	2	14.86	13.9	14.39	14.87	15.33	13.91	14.97	15.81	14.02	14.62		14.6680
	3												#DIV/0!
	Average	14.8	14.74	14.45	14.685	14.44	14.745	14.62	14.98	14.855	14.44	X_a	14.6755
	Range	0.12	1.68	0.12	0.37	1.78	1.67	0.7	1.66	1.67	0.36	R_a	1.0130
B	1	15.33	15.68	15.1	15.09	15.21	15.09	15.21	15.33	15.56	15.33		15.2930
	2	14.97	15.45	15.09	14.86	14.97	14.86	15.45	15.21	15.33	15.09		15.1280
	3												#DIV/0!
	Average	15.15	15.565	15.095	14.975	15.09	14.975	15.33	15.27	15.445	15.21	X_b	15.2105
	Range	0.36	0.23	0.01	0.23	0.24	0.23	0.24	0.12	0.23	0.24	R_b	0.2130
C	1												#DIV/0!
	2												#DIV/0!
	3												#DIV/0!
	Average	#DIV/0!	#DIV/0!	#DIV/0!	#DIV/0!	#DIV/0!	#DIV/0!	#DIV/0!	#DIV/0!	#DIV/0!	#DIV/0!	X_c	#DIV/0!
	Range	0	0	0	0	0	0	0	0	0	0	R_c	0.0000
Part Average (X_p)		14.9750	15.1525	14.7725	14.8300	14.7650	14.8600	14.9750	15.1250	15.1500	14.8250	X double bar	14.9430
												R_p	0.3875
												R double bar	0.6130
												X diff	0.5350
												UCL_r	2.0045
												LCL_r	0.0000

FIGURE 5.22. Gage R&R study worksheet—data.

213

DATA ANALYSIS		
K1 (2 trials = 4.56, 3 trials = 3.05)	4.56	
K2 (2 appraisers = 3.65, 3 appraisers = 2.7)	3.65	
K3 (see MSA page 58)	1.62	
Number of parts (n)	10	
Number of trials - r	2	
	Measurement Unit Analysis	**% Total Variation**
Equipment Variation (EV)	2.80	
EV square/nr	0.3907	% EV 11.46
Appraiser Variation (AV)		
(X diff x K2) square	3.8132	
X diff x K2 square - EV square/nr	3.4226	
AV	1.85	%AV 54.25
R&R	3.352	% R&R 98.29
Part Variation (PV)	0.6277	% PV 18.41
Total Variation (TV)	3.4103	

FIGURE 5.23. Gage R&R study worksheet—data analysis.

Total variation = repeatability + reproducibility

+ part variation + within-part variation

Or

$$TV = \sqrt{(EV^2 + AV^2 + PV^2 + WIV^2)}$$

$$R\&R = \sqrt{(EV^2 + AV^2)}$$

where EV = equipment variation or repeatability
AV = appraiser or operator variation or reproducibility
PV = part variation
WIV = within-part variation

EV, AV, R&R, PV, and WIV are evaluated with respect to TV.

Another approach to the analysis of Gage R&R data would be to use commercially available statistical analysis software such as Minitab, JMP, or Statgraphics. Just enter the data and select the analysis method to determine the components of variation. These software programs also graphically display the data at a click of key. For example, the printout shown in Fig. 5.24 depicts components of variation in a measurement system and variation within each component.

The results of the measurement system analysis for repeatability and reproducibility must be evaluated using the following guidelines:

- If R&R variation is less than 10 percent of total variation, the measurement system can be considered acceptable for all dimensions.

- If R&R variation is 10 to 30 percent of total variation, the measurement system can be considered acceptable for non-critical applications.

- If R&R variation is more than 30 percent of total variation, the measurement system must be considered unacceptable and corrected before use.

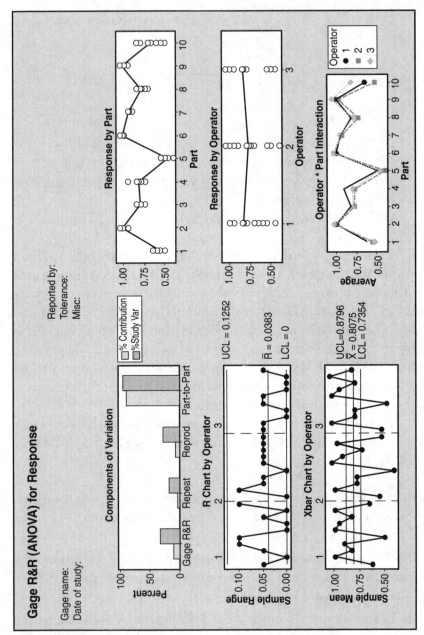

FIGURE 5.24. Gage R&R study using software. (*Printed with permission of Minitab Inc.*)

MEASURES OF SIX SIGMA

In earlier sections we discussed continuous and discrete data. These are important now as we discuss the measurement of process performance, because the process data we get could be discrete or continuous. Each requires a different method of measurement.

Measurements that can assume any possible value are termed continuous data—for example, the customer service cycle time. Time could take any value; the only restriction is our measurement capability. When we say it took 3 minutes, in reality it could have been 3.00056 minutes. Continuous data can take infinite values within a given range.

Measures that assume only limited values are termed discrete data—for example, daily data on the number of times a bank took more than 3 minutes to service a customer. These data can only assume discrete values—2, 4, 100, 1,000, 150,000, etc. Although this value can assume any number, large or small, it will always be discrete—that is, separated by a unit of 1.

The preceding examples also show that both discrete and continuous data can be obtained for the same measure. In the bank customer service cycle time example, individual cycle time data are continuous, whereas daily data on the number of times the cycle time exceeded 3 minutes are discrete. However, continuous data, where possible, are always better than discrete data. You can calculate magnitude of variation with continuous data, and fewer sample data is required.

CALCULATING SIGMA LEVEL

Before we venture into calculating process performance, let's quickly review what Six Sigma means.

Six Sigma is a process performance level, where

- The mean is preferably centered between the customer upper and lower specifications.
- The variation is so small, that 12 times the standard deviation falls within the customer upper and lower specifications.

Figure 5.25 illustrates the Six Sigma concept. Accordingly, the width of the bell-shaped curve demonstrates the process capability, which is 6 times standard deviation. For a process to be at the Six Sigma level, the process capability must be half the tolerance allowed by the customer. In other words, the standard deviation must be such that specification limits are 6 SD away from the process mean. The probability of having a process output outside the specification limits that are 6 SD away from the process mean is 2 parts per billion (ppb). However, according to the Six Sigma methodology, a shift of process mean on either side is allowed to account for uncontrolled variation. If that happens, the probability of having a process output outside the specification limits becomes 3.4 ppm.

For variable data, the probability (in ppm) of a process output or the data point being outside a limit is calculated, and the sigma level is determined.

CALCULATING SIGMA FOR DISCRETE DATA

To calculate sigma for discrete data, you have to know the following three items about what you are measuring.

- *Unit.* A discrete output of a process is called a *unit*. For manufacturing operations, a unit is a product; for nonmanufacturing operations, it could be a report or communication; and for a service business or function, it could be an experience within the scope of the process.
- *Defect.* This is any characteristic that does not meet customer requirements. It is a quantity of product, material, or service forming a cohesive entity on which a measurement or observation may be made.
- *Opportunity for error.* This is an action, item, or event that provides a chance of not meeting the customer requirement, or of producing a defect or an error.

Let's examine the bank customer service example. One of the requirements was that the customer must be served in less

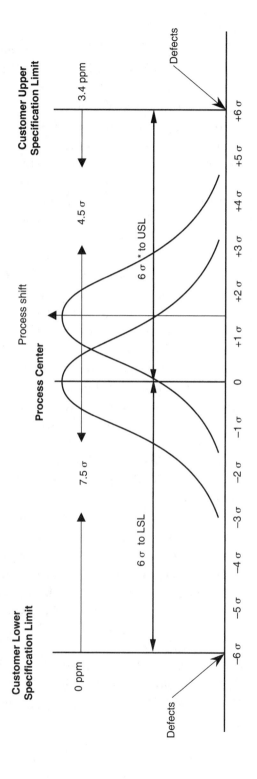

FIGURE 5.25. Six Sigma process data distribution. Total defective ppm for a process with the 1.5σ shift is 3.4 ppm.

than 4.5 minutes (upper specification limit). Let's say the bank is also concerned about customer satisfaction with the service it provides and it asks customers to complete a short survey form where they rate their satisfaction on a scale of 1 to 5 (in increasing order). Any rating less than 3 is considered to be not meeting the customer service requirement.

In some cases, the number of opportunities is obvious, whereas in others it may not be so. The way to determine the number of opportunities is to identify the most critical measures from your customer's perspective where your failure will leave your customer very dissatisfied.

In this example, the customer being served is a unit. The number of items on the survey reflects the number of opportunities for error. There may be additional opportunities for error in delivering the service that is being measured by the survey. Calculation of the number of opportunities for error is the most controversial aspect of calculating Six Sigma. Some people count everything that could go wrong in the process or in a product. Before you determine number of opportunities for error, you must understand the intent of the measurement. The purpose of the measurement is to identify areas for improvement. Therefore, the opportunities should be the areas that can *practically* go wrong while performing various tasks, or the parts that have a *practical* likelihood of malfunctioning. Ultimately, the goal is to reduce opportunities for error in order to eliminate errors from the process or the product.

For example, in the invoicing process, the opportunities could be the number of entries that are made on the invoice. Similarly, in a manufactured product, the number of opportunities could be the number of parts that are being worked on.

The unit can be considered a product-related measurement, while the opportunities for error can be considered a process-related measurement.

To determine the sigma level, you calculate the defects per million opportunities (DPMO), then convert it to the sigma level using a table such as is shown in Fig. 5.26, or using off-the-shelf software.

DPMO	Sigma	C_{pk}	DPMO	Sigma	C_{pk}
3.4	6.0	2.00	66,807	3.0	1.00
5	5.9	1.97	80,757	2.9	0.97
9	5.8	1.93	96,801	2.8	0.93
13	5.7	1.90	115,070	2.7	0.90
21	5.6	1.87	135,666	2.6	0.87
32	5.5	1.83	158,655	2.5	0.83
48	5.4	1.80	184,060	2.4	0.80
72	5.3	1.77	211,855	2.3	0.77
108	5.2	1.73	241,964	2.2	0.73
159	5.1	1.70	274,253	2.1	0.70
233	5.0	1.67	308,538	2.0	0.67
337	4.9	1.63	344,578	1.9	0.63
483	4.8	1.60	382,089	1.8	0.60
687	4.7	1.57	420,740	1.7	0.57
968	4.6	1.53	460,172	1.6	0.53
1,350	4.5	1.50	500,000	1.5	0.50
1,866	4.4	1.47	539,828	1.4	0.47
2,555	4.3	1.43	579,260	1.3	0.43
3,467	4.2	1.40	617,911	1.2	0.40
4,661	4.1	1.37	655,422	1.2	0.37
6,210	4.0	1.33	691,462	1.0	0.33
8,198	3.9	1.30	725,747	0.9	0.30
10,724	3.8	1.27	758,036	0.8	0.27
13,903	3.7	1.23	788,145	0.7	0.23
17,864	3.6	1.20	815,940	0.6	0.20
22,750	3.5	1.17	841,345	0.5	0.17
28,716	3.4	1.13	864,334	0.4	0.13
35,930	3.3	1.10	884,930	0.3	0.10
44,565	3.2	1.07	903,199	0.2	0.07
54,799	3.1	1.03	919,243	0.1	0.03

FIGURE 5.26. Six Sigma conversion table.

$$DPMO = \frac{Defects \times 1,000,000}{units \times average\ opportunities\ per\ unit}$$

Suppose the bank surveyed 100 customers (units) using a 5-question survey, and 10 customers answered a total of 20 questions as unacceptable (i.e., less than 3). Therefore

$$\text{DPMO} = \frac{20 \times 1{,}000{,}000}{100 \times 5} = 40{,}000$$

From Fig. 5.26, we can see that for a DPMO of 40,000, the corresponding sigma value is roughly 3.25.

Another way to calculate the sigma level is to first determine the defects per unit (DPU), then calculate DPMO and determine the sigma level.

$$\text{DPU} = \frac{\text{total defects}}{\text{total units}}$$

Then

$$\text{DPMO} = \frac{\text{DPU} \times 1{,}000{,}000}{\text{opportunities for error in a unit}}$$

To determine the first-pass yield (FPY; i.e., the yield of a process without any repair or rework) the following formula is used:

$$\text{FPY} = e^{\text{DPU}} \qquad \text{Excel command ``= EXP(-DPU)''}$$

Example.   DPU for bank customer service:

$$\text{Defects} = 20 \qquad \text{units} = 100$$

$$\text{DPU} = \frac{20}{100} = 0.20;$$

$$\text{FPY} = e^{-0.2} = 0.81873$$

$$\text{DPMO} = \frac{0.2 \times 1{,}000{,}000}{5} = 40{,}000$$

$$\sigma = {\sim}3.25 \qquad \text{(see Fig. 5.26)}$$

DPU is used to measure the performance of the process output going to the customer or the next process; DPMO is used to measure the performance of a process to produce the desired process output, or the product. Typically, DPU can be used to view the product from customer's perspective, while DPMO can be used to view the process internally so opportunities for improvement can be identified and remedied. DPMO is also a good measure to compare products or process of varying complexity.

PROCESS PERFORMANCE MEASURES

▪ Process capability index C_p
▪ Process capability index adjusted for process shift C_{pk}

Process capability index C_p is the ratio of process tolerance to the process capability, that is:

$$C_p = \frac{\text{upper specification limit} - \text{lower specification limit}}{6 \times \text{process SD}}$$

C_p is an excellent measure to gain confidence in the initial capability to produce the product or service. It can be used to compare the production capability to the design tolerance. It can also be used to evaluate the expected versus the actual. C_p is also known as the *inherent* or *designed-in* capability of the product or service.

If the tolerance of a process is 20 inches, and the standard deviation is 0.5 inches, C_p will equal 20/6 × 0.5 or 6.67. This is an extremely good number. If the width of a car is 8 feet, and the width of the garage door is 12 feet, C_p will equal 12/8 or 1.5 (given that the driver drives in a straight line, as typically occurs when parking the car). Under normal driving conditions, if the driver veers off about 2 feet on either side of the car while driving, the total driving capability would become approximately 2 + 8 + 2, or 12 feet. If the lane on a road is 12 feet wide, C_p will equal 1.0, a borderline case. Any greater deviation or a narrow road would make driving a challenge.

For the process shown in Fig. 5.27, the process capability index for cycle time = (4.5 − 1.5)/(6 × 0.6) = 0.83. This happens to be less than 1.0, implying that a certain percentage of the process output would not be meeting customer requirements for on-time delivery.

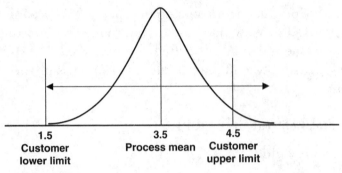

FIGURE 5.27.   Capability index for cycle time (SD = 0.6).

While using C_p, one can note that C_p calculations only consider variance, not the mean or the target value. C_p is a good measure of the process capability with respect to the requirements. However, it does not take into account circumstances in which the process mean shifts to the left or right due to variation. Another measurement, C_{pk}, is used to evaluate the actual process performance. C_{pk} is the process capability index adjusted for shift in the process mean. C_{pk} gives a snapshot of a process at a given time. If C_p is the inherent process capability, C_{pk} is the actual process performance.

There are two formulas for C_{pk}—one is for when the process center is closer to the upper specification, and the other is for when the process is closer to the lower specification.

$$CPU = \frac{\text{upper specification limit} - \text{process mean}}{3 \times SD}$$

Or

$$CPL = \frac{\text{process mean} - \text{lower specification limit}}{3 \times SD}$$

$$C_{pk} = \min (CPU \text{ or } CPL)$$

In the cycle time example, the process mean 3.5 is closer to upper specification 4.5; hence, we will use the first of the two formulas.

$$C_{pk} = \frac{4.5 - 3.5}{3 \times 0.6} = 0.55$$

Here, the value of C_{pk} is less than 1. Therefore, the process will be considered unacceptable, as it will produce a significant percentage of dissatisfied customers. When the process is centered (i.e., there is no shift in the process mean), $C_p = C_{pk}$; otherwise, C_{pk} will always be less than C_p.

The following guidelines can be used to evaluate C_p values:

$$C_p > 1.33 \quad \text{(capable)}$$

$$C_p = 1.00 - 1.33 \text{ (capable with tight control)}$$

$$C_p < 1.00 \text{ (incapable)}$$

And

$$C_{pk} = 1.5\text{–}2.0 \text{ is a respectable value}$$

$$C_{pk} > 1.0 \text{ means } 6\sigma \text{ spread is inside specification limits}$$

$$C_{pk} < 1.0 \text{ means some part of the distribution is outside the specification limits}$$

Figure 5.28 shows examples of varying C_p and C_{pk}. One can reflect on them to comprehend the use of C_p and C_{pk}.

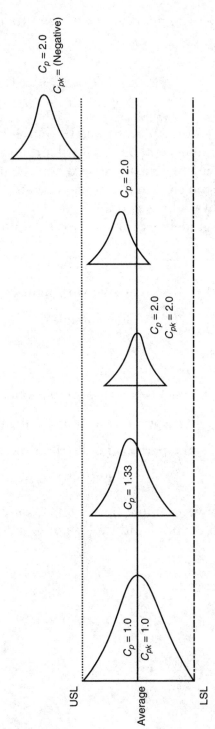

FIGURE 5.28. Examples of varying C_p and C_{pk}.

226

SIX

DEVELOPING A SOLUTION—ANALYZE PHASE

Clearly defining the problem and establishing a baseline of performance are the initial steps in understanding the nature and extent of a problem. Knowing the current state of a process is a prerequisite to establishing the goal or the future state. Many problems are solved by the time a problem is clearly defined and measured. As someone has said, a well-defined problem is half solved. When problems do not reveal themselves after initial definition and quantification, the Analyze phase is pursued. To solve a problem, one needs to perform root-cause analysis, anticipate any problems that could prevent effective implementation, perform regression analysis to establish the relationship between inputs and causes and effects, and perform hypothesis testing to evaluate the new processes. There are many tools that can be used for root-cause analysis and solution development. The following list presents the set of tools most often used to solve most problems:

1. *Fishing for the causes.* Cause-and-effect analysis, also known as *fishbone* or *Ishikawa diagramming,* is an excellent tool for exploring potential causes that could affect the problem. The team working on further convergence or potential solutions then prioritizes the causes.

2. *Looking for trouble.* Failure mode and effects analysis (FMEA) is a great tool for anticipating potential problems so some problems may be prevented.

3. *Variability is evil.*   Multivariable analysis is a tool for apportioning variation into its components, reducing the scope of the problem to a manageable level.

4. *Predicting performance.*   Regression analysis is used to build process models, or to quantify or prioritize the relationship between various causes and effects.

5. *Evaluating change through the power of testing.*   Process improvement requires process changes. Hypothesis testing or comparative tests are used to evaluate the process changes.

6. *Evaluating means.*   Analysis of variance (ANOVA) extends the testing of planned process changes for more than three variables or treatments.

FISHING FOR THE CAUSES—CAUSE-AND-EFFECT ANALYSIS

As someone has said, those who succeed in problem solving in quality control are those who perform useful cause-and-effect analysis. Cause-and-effect analysis is a tool for identifying potential causes of a given problem. It is easy to use. If implemented correctly, cause-and-effect analysis can save significant resources by collectively identifying potential causes. A well done cause-and-effect analysis can solve the problem during the Analyze phase, thus saving further effort in the Improvement phase.

Kaoru Ishikawa, president of the Musashi Institute of Technology in Tokyo, first utilized a diagram similar to cause-and-effect analysis. Therefore, the tool is sometimes called an *Ishikawa diagram* or a *fishbone diagram* because of its resemblance to the skeleton of a fish.

The benefits of cause-and-effect analysis include the identification of potential causes, increased understanding of process issues, delineation of the importance of various process variables, and heightened awareness of the process yields. The graphic nature of the diagram allows teams to organize large amounts of information about a problem and to pinpoint possible causes. Specifically, the cause and effect diagram has the following advantages:

1. It lists the potential causes of a problem.
2. It shows the relationship between the causes and the problem.
3. It uses cross-functional teamwork to improve understanding of the process.
4. It sometimes identifies the root cause of a problem.
5. It is simple to use and easy to understand.

The cause-and-effect analysis or fishbone diagram segregates potential causes in four to six major categories. The four main categories are *material, method, machine,* and *people* (people power or mental power). Two other categories some users like to add are *measurements* and *environment.* However, if the team feels a need for broader categorization, more categories are added. Employees representing various departments such as production, maintenance, quality, purchasing, and calibration identify the potential causes in a brainstorming session. The team members can identify causes for each branch, either separately or together.

Successful analysis requires a group leader to facilitate the brainstorming session. The method essentially identifies all potential causes of variations or effects that may contribute to a problem or a cause. Taking a process view of work performed, we classify causes of variations into the following categories:

1. *Input variations.*   Materials, tools, information, requirements, procedure, equipment condition, setup conditions, and people skills.
2. *Process variations.*   Machine maintenance, equipment-mounted or stand-alone monitoring devices; compliance to procedures or methods; material consistency; operator discipline, consistency, and data collection; software version or errors; and environmental conditions.

3. *Output variations.* Verification method, inspector skills and consistency; calibration of measuring or verification devices; product design quality; product performance parameters; workmanship standards; and methods to handle nonconforming material.

A sample cause-and-effect diagram is shown in Fig. 6.1. The effect or the problem statement is written in the head of the fish, or the box. Major branches or bones are attached to the head. Twigs or subbranches are attached to the major branches as a potential cause is further explored for an actionable, verifiable, and measurable cause.

Once the potential causes are listed, the team prioritizes the causes, either by branch or overall for the problem as a whole. When team members struggle in selecting important causes, each member must be required to select at least one cause on each branch that is considered the most critical, order rank the causes, and identify the root cause or the major causes. While listing causes at a lower level, each potential cause must be questioned at least five times to get to the actionable, quantifiable, and verifiable root cause. While searching for a root cause of a complex and chronic problem,

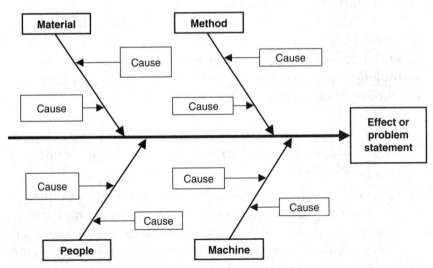

FIGURE 6.1. Cause-and-effect diagram.

one may need to develop a process flowchart, and then conduct the cause-and-effect analysis at each process to identify various causes. In this case, there may be multiple fishbone diagrams. After identifying one or more potential root causes, an action plan is developed to alleviate the root cause of the problem. If there is a consensus about one root cause, then a corrective action is developed to remedy the root cause. When there are multiple potential causes, the next-level techniques are deployed to learn more about the process.

CONSTRUCTING A CAUSE-AND-EFFECT DIAGRAM

To construct a cause-and-effect diagram, the problem must be clearly defined and the team members must be identified. The team leader is designated to facilitate the team exercise, and members agree to participate. Then a team meeting is set up where the basic rules of equal participation apply; that is, each member gets a chance to identify causes without being questioned or criticized. The following lists various steps in constructing the cause-and-effect diagram:

Step 1. *Identify and state the problem to be solved.* The problem is written in a box, and a long arrow, or the backbone, is drawn pointing to the box.

Step 2. *Ensure that the right people are present during the construction of the cause-and-effect chart.* Typically, one must have representatives from the area management, materials or purchasing, maintenance, engineering and quality, production, and other departments as necessary. Arrange the seating in a U-shape or oval in order to increase interaction among the participants.

Step 3. *Construct main branches and label them* People *(people power or mental power),* Material, Method, *and* Machine. Other branches (measurements and environment) can be added if the team deems necessary.

Step 4. *Conduct a brainstorming session to identify potential causes.* During the brainstorming session, any potential cause is just listed without questioning. To explore further, the team

asks at least five *whys* in trying to get to the deeper understanding of the relationship between the cause and the effect.
Each *why* peels away the layer of ambiguity and gets closer to
the root cause of a problem. As causes are identified, subbranches are attached to the main branches. Each cause can
be questioned for availability of data as well. Following is a set
of guidelines for the brainstorming session:

- *Team members do not evaluate or criticize others' ideas during
  the process.*   Criticism of others' ideas will only shut off the
  flow of ideas.
- *Everybody, irrespective of rank and position, should be given
  equal air time during the brainstorming session.*   Senior or
  dominating members of the team should not monopolize the
  discussion, and should not dominate the discussion.
- *Get as many ideas as possible.*   At the beginning, you don't
  need to worry about the quality of ideas. Encourage quantity
  over quality in the early stage of the process. Each member
  should first be asked to contribute one idea before the second round of ideas, and so on.
- *Encourage piggybacking.*   That is, one member of the team
  could get ideas from other team members and extend those
  ideas to new depths. Generally, one member's idea becomes
  a launching pad for the other ideas.
- *Record every idea.*

Step 5. *Ask why this problem occurs and write down the
answer below the problem.* Have team members write down
their ideas on 3- × 5-inch cards or loose sheets of paper before
starting the discussion. Once all the causes are listed, the team
identifies the most likely causes. At this stage, the team discusses various causes and evaluates the significance of each
based on their process knowledge. Sometimes teams cannot
make a decision soon enough; they end up looking at higher-
level causes, which dilutes the team activity. The best way to
quickly decide which are the most likely causes is to ask the
team to start removing less likely causes. This narrows down

the list of potential causes to more likely causes. The following are probing questions about key factors to gain insight on them:

1. Is this cause a variable or an attribute?
2. Has the cause been operationally defined?
3. Is there a control chart or other data available?
4. How does this cause interact with other causes?

Step 6. *Prioritize the most probable cause for corrective action.* Prioritization is based on the importance of the most likely causes, available resources, and the ease of verification of causes. The objective is to identify at least one cause on each major branch with team consensus. By selecting one critical cause per branch, the team reduces the number of variables from approximately 25 to 4.

Step 7. After the most probable causes are identified, the team must decide which one cause is most likely the root cause. If the team members can agree on the root cause, the exercise moves into the solution development phase.

After the root cause has been identified, an appropriate corrective action is identified to remedy the problem. The remedial actions must be validated to ensure the effectiveness of the corrective action.

During the brainstorming session, the team leader plays an important role in constructing the cause-and-effect diagram. The success of the cause-and-effect analysis depends on the experience and participation of its team members. A well- constructed cause-and-effect diagram, and a well-run brainstorming session, makes team members feel happy and creates a sense of breakthrough or turning on the light.

Figure 6.2 shows a cause-and-effect diagram for the plating process. The problem or the effect is excessive variation in plating thickness. Start the process by drawing the fishbone and drawing the four major branches—material, people, method, and machine. Structured brainstorming sessions will bring out likely causes and help the team complete all four branches of

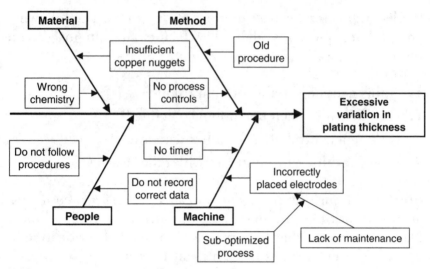

FIGURE 6.2.   Cause-and-effect diagram of plating thickness problem.

the diagram. After identifying all possible causes, the team needs to rank the causes and identify the most likely cause.

In another example, shown in Fig. 6.3, the problem is excessive network time. Start the process by drawing the fishbone. At the end of the horizontal line or in the head of the fish, write the problem down. Next, draw four major branches, including hardware problems (equipment), people, software problems (information), and outside problems (method). Successful identification of the bones helps in constructing the fishbone diagram. Structured brainstorming sessions will bring out likely causes and help the team complete all the branches. After identifying all possible causes, the team needs to rank all likely causes and identify the most likely causes. Some causes initially appear to be effects of some other causes. Further questioning leads to the root cause. For example, network sabotage by a disgruntled employee initially appears to be the cause of the problem. However, further investigation may lead to such causes as unfair compensation, mistreatment of employees, and the like.

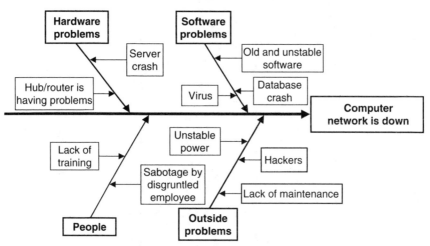

FIGURE 6.3. Cause-and-effect diagram of computer network problem.

LOOKING FOR TROUBLE—FMEA

Failure mode and effects analysis (FMEA) is a systematic method, as shown in Fig. 6.4, of identifying potential failures with the objective of preventing their occurrence to minimize the probability that the customer will see a failure. The potential failures are those that customers will perceive as a failure. Here the customers are both internal and external. In performing FMEA, one attempts to identify each potential mode of failure, its effect, and its severity and address its causes. After the causes have been identified, the failure mode and its effects are mitigated through corrective actions. FMEA is normally conducted while developing a solution for the problem, minimizing surprises during the implementation. Normally, FMEA is deployed in the product or process design phase.

In essence, FMEA is a risk-minimization technique that helps the organization identify a risk with a proposed solution in a Six Sigma project, investigate its root cause, and initiate efforts to reduce the risk. The technique was first used in the aerospace industry in the mid-1960s to find problems with an aircraft before it ever left the ground. The technique has since been used in many industries, such as the aerospace,

Item and Function	Potential Failure Mode	Potential Effects of Failure	Potential Causes of Failure	Detection Methods and Quality Control	DET	RPN = SEV × PROB × DET	Recommended Action
Part name and function	Possible modes of failure	Consequences of failure on part function and on the next assembly	Causes such as inadequate design and improper materials	Measures available to detect failure before they reach the customers		Total RPN	List actions for each failure mode identified as significant by the RPN rating

FIGURE 6.4. FEMA: analysis of potential failures.

automotive, and pharmaceutical fields. Applying FMEA when using the DMAIC method to improve the profitability of a company is as critical as it is in matters of human safety. FMEA offers the following advantages:

- It identifies potential failure modes that could have been overlooked and makes the solution more effective.
- It ranks potential failure modes based on severity, occurrence, and detection. The ranking allows the prioritization of failure modes and monitors the effectiveness of the solution while recalculating the ranking.
- Continual application of the FMEA process during the product or process life cycle leads to better quality and reliability of products, with greater customer satisfaction.
- FMEA creates a document that contains a significant amount of knowledge about a product and process. FMEAs can contribute to the knowledge management effort to develop innovative solutions.

An important factor for the successful implementation of an FMEA program is *timeliness*. FMEA is more effective when it is proactively applied in the design phase of a process, product, or service. Services involve intangibles and perceptions more so than in products. Customer participation in the service process makes it even more challenging. Considering customers also includes internal functions such as design, engineering, marketing, finance, and sales. The FMEA technique can help in anticipating and reducing potential errors passed on to the internal customers.

FMEA can be used in the evaluation of design concepts, process selection and improvement, design process analysis, software development, and services. FMEA methodology consists of grading failure modes for severity, potential causes for occurrence, and controls for detection. The definition of these terms may vary across companies and products. Therefore, one must standardize the definitions before applying FMEA for consistency and accuracy.

Severity. This involves the seriousness of the effect of the potential failure mode on the functionality of the product or on customer applications. It gives an idea of the resulting loss generated directly or by adverse effects on the subsequent operational steps. The severity is graded on a scale of 1 to 10, where 10 relates to a life-threatening situation, and 1 implies minimal effect.

Occurrence. This represents the frequency with which a cause of potential failure may occur. The frequency of occurrence is estimated based on the process knowledge and the historical performance. In the absence of historical data or knowledge, the frequency of occurrence may be estimated based on similar processes, or as determined by the cross-functional team. The occurrence is graded on a scale of 1 to 10, where 10 implies certainty of an event, and 1 implies a hypothetical situation.

Detection. This represents the relative probability with which the effect of a cause can be detected through appropriate controls (i.e., inspection, test, or process control). Detection is also graded on a scale of 1 to 10, where 10 implies difficulty in detection, and 1 implies a certain containment of adversely affected material.

Performing FMEA

In conducting an FMEA while working on a Six Sigma project, the proposed solution or the process change is evaluated for potential problems or challenges that could prevent the desired results from occurring. Following are the steps in performing an FMEA:

1. Identify the project and understand all components and their functions.
2. Select a cross-functional team from all affected work groups that could contribute to the completion of the FMEA.
3. Gain consensus on the ranking criteria for severity, occurrence, and detection. Figure 6.5 shows the typically used guidelines.

Severity	Severity Ranking	Probability of Failure	Ranking	Detection	Detection Ranking
Hazardous without warning	10	Very high: failure is almost inevitable (>1 in 2)	10	Absolute uncertainty	10
Hazardous with warning	9	1 in 3	9	Very remote	9
Very high	8	High: repeated failures (1 in 8)	8	Remote	8
High	7	1 in 20	7	Very low	7
Moderate	6	Moderate: Occasional failures (1 in 80)	6	Low	6
Low	5	1 in 400	5	Moderate	5
Very low	4	1 in 2,000	4	Moderately high	4
Minor	3	Low: relatively few failures (1 in 15,000)	3	High	3
Very minor	2	1 in 150,000	2	Very high	2
None	1	Remote: failure is unlikely (<1 in 1.5 million)	1	Almost certain	1

FIGURE 6.5. FEMA: Ranking of potential failures.

4. Draw the process flowcharts and describe the functions of each component.

5. Identify necessary inputs to the process, such as materials, method, machines, and people actions. Anticipate and list all potential failure modes for each input at various steps in the process. Team members can brainstorm for ideas and potential failure modes based on their experience and process knowledge.

6. Evaluate the severity of the effect of the potential failure mode. Consider the degree of cosmetic, function, and safety aspects in estimating the severity of the effect. The effect can be divided into catastrophic, critical, major, minor, and negligible categories. Accordingly, assign the severity ranking on a scale of 1 to 10.

7. Investigate potential causes of the failure mode. If causes are difficult to identify, construct a cause-and-effect diagram. Evaluate the frequency of occurrence and assign the ranking on a scale of 1 to 10.

8. Calculate the risk priority number (RPN) by multiplying severity S, occurrence O, and detection D as follows:

$$RPN = S \times O \times D$$

9. The maximum possible RPN value is 1,000. RPN provides a relative ranking of the causes of failure modes. Sort them by RPN in descending order. Based on the severity, RPN, and available resources, establish an action plan to reduce RPN. An action can be triggered for RPN as long as it makes economic sense.

10. Establish a threshold value of RPN for taking action. If the severity is greater than 7, some action must be initiated to address the effect.

11. Identify ways to reduce the risk of the highest priority failure mode by using poke-yoke and error-proofing methods.

12. Develop a plan to reduce RPN and address necessary failure modes by addressing the following actions:

- Assign responsibilities for further action.
- Outline an action plan to reduce or eliminate failure mode.
- Implement corrective action and reassess RPN.
- Continue until all issues are addressed.

Figure 6.6 shows a sample FMEA form that follows the preceding steps. Figure 6.7 shows an example concerning the seat belt installation process at an automobile plant. Various defects could occur while installing the seat belt. Only three defects are listed here for illustrative purposes. You need to identify the severity of each defect on a scale of 1 to 10. Leaving the seat belt loose will likely be the most severe defect in this case, hence a ranking of 9. The frequency of occurrence and ability to detect are low. Therefore, the overall RPN may not be significant. In such cases, when the RPN for a failure mode is not significant, but severity is, the failure mode must be addressed. Otherwise, the failure mode with the highest RPN must get the highest priority for process improvement.

FMEA is increasingly being used in service sector. For example, a physical therapy clinic is interested in performing

FEMA number ________________

Item ________________ Process responsibility ________ Prepared by ________________

Model year(s)/ vehicle (s) ______ Key date ________________ FMEA date (orig.) ______ (rev.) ______

Core team ____

Process Function Requirements	Potential Failure Mode	Potential Effects of Failure	SEV	Potential Causes/ Mechanisms of Failure	OCCUR	Current Process Controls	DETEC	RPN	Recommended Actions	Responsibility and Target Completion Date	Action Results				
											Action Taken	SEV	OCC	DET	RPN

FIGURE 6.6. Sample process FEMA.

Failure Mode and Effects Analysis
Process Name: Seat belt installation

Date: _____

Process number: _____

Failure Mode	Severity S	Probability of Occurrence O	Probability of Detection D	Risk Preference Number RPN $= S \times O \times D$
1. Selecting wrong seat belt	6	3	4	72
2. Seat belt left loose	9	2	7	126
3. Cover clip not aligned	3	3	5	45

FIGURE 6.7. FEMA: Seat belt installation.

FMEA. A patient makes an appointment to meet a physical therapist. Figure 6.8 shows a list of various failure modes or errors that could occur during the therapy process. Once failure modes are identified, severity and detection are assessed on a scale of 1 to 10.

It appears that improper treatment given to a patient has the highest RPN, and is therefore a likely candidate for process improvement.

VARIABILITY IS EVIL—MULTIVARIABLE ANALYSIS

A process problem normally occurs due to variations in *material* (information), *method* (approach), *machine* (tools), and *mind power* (employees' intellectual involvement). The uncontrolled variation is sometimes difficult to understand and pinpoint, as there are so many variables involved that could be the cause. One can analyze production data, using Pareto charts, histograms, and other tools. However, it is difficult to pinpoint the root cause. For processes and problems that are complex, multivariable analysis is a tool to classify variation into three smaller chunks, allowing one to divide and conquer the problem. Multivariable analysis divides the total variation into three families of variation. The family with the most variation is attacked for reduction. The multivariable chart is a graphical representation of variation that helps to visually identify the major family of variation. The three families of variation are the following:

1. Positional variation
2. Cyclical variation
3. Temporal variation

Figure 6.9 shows multivariable analysis for a plating process. In the plating process, three families of variation, *positional, cyclical,* and *temporal,* are defined as variation within a plating panel, between batches, and over time, respectively.

FAILURE MODE AND EFFECTS ANALYSIS
PROCESS NAME: PATIENT RECEIVING PHYSICAL THERAPY

DATE: _____

PROCESS NUMBER: _____

FAILURE MODE	SEVERITY S	PROBABILITY OF OCCURRENCE O	PROBABILITY OF DETECTION D	RISK PREFERENCE NUMBER RPN $= S \times O \times D$
1. Patient missing appointment	2	3	7	42
2. Patient providing confusing information about the symptoms	5	2	6	60
3. Improper treatment given to a patient	9	2	5	90
4. Patient fails to understand physical therapy benefits	4	2	4	32

FIGURE 6.8. FEMA: Physical therapy at the clinic.

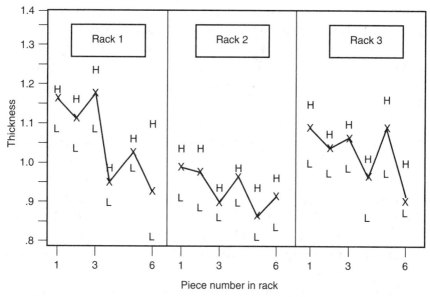

FIGURE 6.9. Multivariable analysis for the plating process: L = low; X = average; H = high.

The chart shows the data for three consecutive batches and panels within each batch. This allows one to review positional and cyclical variation. If the data for more batches had been plotted, that might have demonstrated temporal variation as well. By plotting the data, one can easily see the largest range of variation in each family and then select the one with most variation to work on.

The effective use of multivariable analysis depends on good planning for data collection, some knowledge of the process, knowledge of defects or variation, and the ability to investigate the process for sources of variation. Multivariable charts allow one to visualize a process problem from a new perspective. Multivariable charting has not been used enough, even though it is a powerful tool. Using a multivariable chart is like solving a puzzle, because once you know the major family of variation, investigating the process from that perspective leads you to a quick solution.

To use multivariable analysis, one must believe that variation is the source of all problems. Therefore, reducing variation is the way to solve process problems. The mindset must be on solving process problems instead of product or part problems. To revisit the variation type, one must understand the difference between common or *random* variation and special or *assignable* variation. Common causes generally account for 80 to 95 percent of the observed variation in a process and generally involve natural variation present in a process. These variations are present in the system due to inherent, built-in, uncontrolled causes. Better training, improved design, better process controls, and better technology can reduce these variations. On the other hand, the special or assignable variations are introduced due to some circumstance. The amount of variation introduced due to assignable causes is relatively large. One way to look at assignable causes is that the chance of such an event happening is small; however, when it happens, it causes a lot of shift in the process.

One must recognize that people tend to create unwanted variation by tampering with a process that is in statistical control, or often by indiscriminately trying to remove common causes of variation. In other words, too much process control is not good either.

PERFORMING MULTIVARIABLE ANALYSIS

To use multivariable analysis, one must define the problem in terms of *what, where,* and *how much.* It is even better if the problem can be quantified using variable data. Then one defines three families of variation in terms of process operations.

Positional variation implies the variation observed at a certain location of the process output or within a process. Positional variation shows a pattern or repetition. For example, the positional variation on a printed circuit board relates to inconsistencies or defects at certain locations on the board. Therefore, the positional variation could be defined as a variation within the board. To capture the positional variation, the board is divided into quadrants, or six or eight areas, each side of the board, and the edges versus the center of the board, and data are collected to reflect that process. By understanding problem symptoms at a certain location, one can focus on process variables that could create such a variation. For example, if boards are being scratched at a certain location on the board, such as the edges or the center, one can look into the scrubber brush bristles, fixtures, or similar variables at specified processes. Since the defect or variation is repetitive, this could be related to the board design. One must understand that one is studying patterns of variation in order to identify the root cause.

Cyclical variation is defined in terms of the process cycle and is caused by changes in the process setup. Typically, such variation is observed between batches or process cycles. This variation in process can be attributed to a new setup, a change in tools, corrective maintenance, a change in chemistry, or a change in personnel. For example, if the cyclical variation is a significant variation in plating thickness, the plating current set by the operators could be the cause.

Temporal variation is defined as the variation observed over time. This variation is reflected in trends or shift over time. The shift could be due to the preventive maintenance cycle, tools wearing out, variations in optical or laser devices, periodic cleaning, or refills.

Once the family of variations is defined in terms of process variables and conditions, the data collection plan is developed. The data collection plan must clarify the amount of data, the log sheet design for collecting the data, and the analysis of the data. The first thing production management wants to know about is how long it will take to collect the data. No one wants to interrupt operations from products or services.

One should collect enough data to capture most of the positional, cyclical, and temporal process variation. At least several process cycles must be included in data collection and a few temporal iterations (shifts and days). The log sheet must be designed such that data are recorded logically, and organized such that the log sheet can quickly show trends or patterns in the data.

Reducing variation really solves many problems at once rather than trying to fix one product problem at a time. Due to the long list of problems, it would take forever to solve product problems that way. One could reduce variation in service processes too—for example, variation among bank tellers, among sales registers in a department store, in customer service response time, or in waiting time at a doctor's office or clinic.

Analyzing a multivariable chart is relatively easy. Visual analysis can identify the major family of variation. Reviewing the multivariable chart in Fig. 6.8, it appears that pieces 2, 5, and 6 have the most variation in each rack and piece 6 has the most variation within a rack. So the variation reflected by piece 6 in rack 1 is a measure of positional variation. The cyclical variation between each rack (or batch) can be calculated by taking the average of measurements for each rack, then taking the range between highest and lowest average value for each rack. It appears that the positional variation exceeds the cyclical variation. Therefore, one can focus on reducing the positional variation first by focusing on variables that are responsible for such variation.

PREDICTING PERFORMANCE: REGRESSION ANALYSIS

The purpose of regression analysis is to build a model describing the relationship between a dependent variable and several independent variables. Regression involving multiple independent variables is called a *multiple regression*. For example, a defective soldering process is studied for possible causes of defects. Likely causes of defects that in turn affect quality include solder bath temperature, vibration of the wave, preheater temperature, the air knife, the conveyor angle, and the solder wave height. Statistical techniques such as ANOVA will reduce the number of variables that have a significant effect on the soldering defects. One likely conclusion is that only bath temperature affects soldering defects. In this case, bath temperature will be an independent variable and soldering defects will be a dependent variable.

Sometimes, independent variables could be "dummy variables" with a value of 0 or 1, showing the presence or absence of certain characteristics; for example, the use of a certain preheater could be a dummy variable in the preceding case. A dummy or qualitative variable is one that only takes on the values 0 and 1—a value of 0 shows the absence and a value of 1 shows the presence of this effect.

The regression technique is a valuable tool to use after identifying root causes. The regression model quantifies the cause-and-effect relationship. For example, a hospital concerned about the length of time required to get a patient from the emergency department to a bed may identify several potential causes, such as the number of patients in the emergency room, the availability of a nurse, the availability of medical or surgery units, or the readiness of the bed.

An automobile insurance company may want to predict the risk factor for a driver before deciding on the premium. This risk factor (dependent variable) will likely depend on the driver's age and marital status, number of dependent driving children, distance to workplace, type of car, claims filed in the past, incident reports during the past 3 years, and the age of the car, to name a few. Once information on a driver regarding the independent variables has been compiled, the company looks at historical data and determines the risk factor and associated premium. Based on regression analysis, insurance companies have found that the age of a driver, below a certain cutoff, makes a difference in risk due to typical driving styles or patterns.

People deploy regression analysis continually, whether using formal or informal methods of calculation. People estimate the correlation between events they observe around them. In a Six Sigma project, the regression analysis can be used to prioritize independent variables, or establish a causal relationship between output and inputs. Once the relationship or model has been established based on historical data, one could predict the dependent variable for a given independent variable. Regression analysis can be seen as analysis of scatter plots by adding a best-fit line and quantifying the relationship between the dependent and independent variables. The regression analysis includes the regression equation, residual variance, and R^2. The regression equation defines the best-fit line. The *best-fit line* is the line that relates the dependent variables to the independent variables. It produces the smallest difference between the actual and predicted values.

The following aspects of regression analysis must be understood to use regression analysis effectively:

- *Assumption of linearity.*   Before going ahead with the analysis, it's a good idea to look at the scatter plot of two variables to confirm this assumption.

- *Variability in the independent variables.*   Not all the independent variables have equal variation.

- *Spurious correlation.*   An implicit assumption one makes when using regression analysis is that the dependent variable is caused by independent variables. However, the stronger statistical correlation does not represent a causal relationship between the dependent and independent variables. For example, one could correlate rainfall in a particular geographical location to unemployment and try to predict the employment rate based on the amount of rain. This relationship will be a spurious one.

- *Choice of number of variables in building the process model.* It is very tempting to include more variables to make the model look more complete. One needs to test the significance of each variable and ask whether the model will be worse off if this variable is dropped. The balance is between parsimony and completeness. As a rule of thumb, one needs to have at least 20 times as many observations as variables.

- *Multicollinearity.*   This is when two independent variables are either completely or very strongly correlated with each other. For example, assume you have length in meters and width in inches as two variables. These two variables happen to be completely correlated. One must recognize such collinearities.

- *Residuals say it all.*   Residuals are normally distributed and independent. One should produce histograms for the residuals as well as normal probability plots in order to validate the assumptions. The residual plot will also indicate the presence of outliers.
- *Error-related assumptions:*
    Expected error.   Error is a random variable with a mean of zero.
    Homoskedasticity.   Error variable has a constant variance.
    Serial independence.   Error terms are independently distributed.
    Normality of error.   Errors are normally distributed.

Figure 6.10 depicts a model of regression analysis. The purpose of regression analysis must be clearly stated and understood. To make the decision to use regression analysis, one must consider the following questions:

- Is regression the right tool for analysis?
- What is the dependent variable?
- What are independent variables?
- What tool should be used to perform regression analysis?

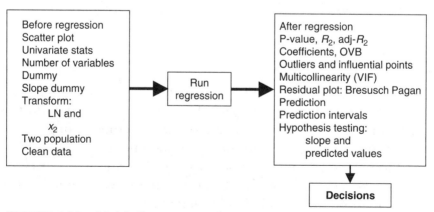

FIGURE 6.10.   Model of regression analysis.

The following is a typical regression model equation:

$$y = \beta_0 + \beta_1 x + \varepsilon$$

The model predicts a dependent variable y, based on the value of the independent variable x. The error term ε (epsilon) is often assumed to be 0 and not written as part of the model. β_0 is the constant or y-intercept and will correspond to a value if the values of all independent variables are 0. β_1 is the coefficient for the independent variable.

Each estimate of the coefficient is subject to sampling error and has a distribution. Regression output provides estimates of standard deviation. R^2 values and adjusted R^2 values, not shown with the equation, tell the percentage variance in y that is explained by knowing x. Adjusted R^2 values will be equal to or lower than R^2 values. Adjusted R^2 values will be lower and lower as the number of independent variables increases.

A natural extension of the analysis considered so far is to have more than one predictor variable X:

$$Y = \beta_0 + \beta_1 x_1 + \beta_2 x_2 + \beta_3 x_3 + \cdots + \varepsilon$$

Essentially, everything that we have said so far about simple linear regression carries through for multiple regressions. The only technical problem is that formulas for the betas and associated variances involve matrix inverses and thus are impractical to apply without software.

SELECTING VARIABLES

Two methods can be used to determine how many and which variables to include in a model. The method used to build the regression model could be the statistical method, using t-ratio and P-value, or the model builder's judgment. The statistical method alone is not a good process for determining or excluding a variable, and should not be the basis recommended for the selection of the model's coefficients.

In modeling product promotions, it is very important to include deal discount, feature advertising, and display activity, because these are the variables that a retailer controls. They should be included in the model if data are available for them. For some purposes, more detailed information might be very useful regarding the nature of the display or the nature of the feature advertisement. For example, knowing the type of display (end of aisle, wall of value, etc.) can often be very useful.

Seasonality and trend are often included in models to control other factors, because omitting them would cause spurious correlation between the independent variables and the dependent variable. Shelf price is often included in a promotional model because it controls for a key factor that may again cause spurious correlations, because promotional effects are tied to changes in shelf price.

The term *parsimonious* is used to indicate that the model should contain only "necessary" variables. While one can always identify other factors that can influence sales and might be included in the model, the best models are those that contain as few variables as possible. Including too many variables in a model results in both "overfitting" and imprecise estimates of the model coefficients. Therefore, it is important to limit the model to critical variables. For example, one would like to include the type of feature advertising in the model rather than simply using the presence of feature advertising. If the type of feature advertising is not going to be used to design the firm's strategies, and the difference in the effects of the different types is small, then there is no reason to include several feature advertising variables (one for each type) in the model. Deciding when a model is not parsimonious is difficult, but the rule should be: do not include a variable unless it is absolutely necessary.

In the real world, it is commonplace to find that data have not been collected for key causal variables. The reason is that the retailer or the data supplier had no reason to collect them, or they are too expensive to collect. Therefore, while a variable may be important, data are not available. The rule is to do the best one can with the data available. A model will never be perfect, and some analysis is better than no analysis.

As an example, regression analysis output is shown in Fig. 6.11. It establishes the relationship between solder defects

PARAMETER	CONSTANT	TEMPERATURE
Coefficient	-1.5974139	0.28705317
Standard error of coefficient	0.37167145	0.02938977
t-Ratio	-4.2979	9.7671
P-value	0.0107%	0.0000%
Standard error of regression		0.58935028
R^2		70.46%
Number of observations		42
Residual degrees of freedom		40
t-Statistic for computing 95% confidence intervals		2.0211

FIGURE 6.11. Regression analysis output: solder defect with bath temperature.

with the solder bath temperature T. The regression model can be written as follows:

$$\text{Solder defects} = -1.59 + 0.28T$$

The equation will be helpful in predicting solder defects within the historical temperature range. The preceding model has an R^2-value of 70.46 percent and a P-value for T, making the independent variable highly significant.

Another example is to determine the effect of customers' income and the presence of competition on the sales of a product. Here we shall consider dummy and slope dummy variables. In this case, competition would be a dummy variable, having a value of 0 or 1, indicating the absence or presence of competition. However, the difference in sales based on competition alone will not be a fixed value. The regression model can be written as follows:

Sales = constant + constant × income + constant

× competition (0 or 1)

The presence of competition will make a fixed adjustment to the model for calculating sales, as shown in Fig. 6.12.

Parameter	Constant	Income	Competitor
Coefficient			
Standard error of coefficient			
t-Ratio			
P-Ratio			

FIGURE 6.12. Regression equation parameters with dummy variable.

However, one could argue that the effect of competition will not be a fixed value, but will depend on the income group of the customers. We need to use a slope dummy variable for this purpose. Slope dummy is the product of a dummy variable (in this case competition) and another variable (income in this case). The revised parameters would be as shown in Fig. 6.13, and the model can be written as follows:

$$\text{Sales} = \text{constant} + \text{constant} \times \text{income} + \text{constant} \times \text{compettion}$$
$$(0 \text{ or } 1) + \text{constant} \times \text{competition} (0 \text{ or } 1) \times \text{income}$$

One needs to check *P*-values for the significance of these variables. The R^2-value will indicate the variation that can be explained with this model.

Performing Regression Analysis

Following are the steps in performing a regression analysis:

1. Decide whether regression is the right tool to quantify the cause-and-effect relationship.
2. Construct a scatter plot for various variable data to check the assumption of linearity.
3. Check for any missing or incomplete data.
4. Identify the dependent variable and select dependent variables.

PARAMETER	CONSTANT	INCOME	COMPETITOR	COMPETITOR × INCOME
Coefficient				
Standard Error of Coefficient				
t-Ratio				
P-Ratio				

FIGURE 6.13.   Regression equation parameters with dummy and slope dummy variables.

5. Run the regression using appropriate software.
6. Check for R^2, P-values (significance), and signs of different coefficients. These three numbers together will indicate the validity and usefulness of the model.
7. Check the plot for residuals and test for normality, independence, and constant variance assumptions using the guidelines shown in Fig. 6.14.
8. Drop any insignificant variable or transform variable if certain assumptions are violated. Rerun the regression.
9. Once you are satisfied with the model, write down the model.
10. Use the preceding model for predictions.
11. Check the model for common sense and statistical and economic significance.

Check the model for validity in terms of signs of the coefficients of independent variables. Ascertain whether the predicted sign of the model makes sense.

For simple analysis between the two variables, one can draw the scatter plot and evaluate the relationship in terms of positive or negative correlations. Calculating the correlation coefficient R and R^2 will establish the strength of the correlation between two variables.

ISSUE OR COMMON ASSUMPTION	SOLUTION TO COMMON VIOLATIONS
Error distribution is normally distributed.	Draw histogram of residuals.
Linearity assumption is confirmed.	Draw scatter plot and plot the residuals.
Error variance is constant.	Plot the residuals versus predicted values.
Errors are independent.	Plot residuals versus time period.
Outliers and influential observations are identified.	Check leverage ratio and studentized residuals.
Multicollinearity problem is reorganized.	Check correlation coefficient and P-values.

FIGURE 6.14.   Assumptions and their validation.

EVALUATING CHANGE THROUGH THE POWER OF TESTING: HYPOTHESIS TESTING

Hypothesis testing is a statistical technique that is used to support an experimental statement. The primary purpose of this technique is to convince stakeholders that the statement is likely true or not true. The technique involves setting up two hypotheses, null and alternative statements, that include all the possibilities and do not overlap. H_a denotes the alternative hypothesis, and H_0 denotes the null hypothesis. The outcome of any hypothesis testing includes the two possibilities:

1. Rejecting the null hypothesis
2. Not rejecting the null hypothesis

The outcome depends on the statistical significance of the evidence. The statistical significance evidence is compared with a threshold value based on the required *level of significance* or confidence. The statistical significance of evidence is called the P-value, expressed in probability terms. As a rule of thumb, low *P*-values will correspond to strong evidence against the null hypothesis, hence supporting the alternative hypothesis. High *P*-value corresponds to weaker evidence.

A criminal trial in a court of law provides a good analogy to aid in understanding this concept. The prosecuting attorney wants to prove to the jury that the defendant is guilty. Likewise, whatever hypothesis we are trying to prove is the alternative hypothesis. In this case, the alternative hypothesis is that the defendant is guilty. All other possibilities, not included in the alternative hypothesis, are included in the null hypothesis. Here, the null hypothesis is that the defendant is not guilty. These two hypotheses should cover all the possibilities and should not overlap. The two possible outcomes of this trial are verdicts of *not guilty* and *guilty*. Figure 6.15 shows the two types of error possible with these outcomes.

The test can never result in the rejection of the alternative hypothesis or, equivalently, in the acceptance of the null hypothesis. If the jury does not reject the null hypothesis that the defendant is not guilty, this simply means that the evidence is not strong enough to prove the defendant guilty. It does not mean that the defendant is innocent. It is very important to make sure when setting up a hypothesis test that what one hopes to prove is stated in the alternative hypothesis.

Hypothesis testing can lead to wrong conclusions, although at a predefined, low probability. There are two types of errors, Type I and Type II. Setting the level of significance is one major step, and the value will depend upon the context. In some contexts where one is looking for very strong evidence, one might set the type error at 1 percent or 0.01. If one is not

OUTCOME	DEFENDANT INNOCENT	DEFENDANT GUILTY
Guilty verdict	Type I error	Correct
Not guilty verdict	Correct	Type II error

FIGURE 6.15.   Type I and Type II errors for the courtroom example.

that concerned about the error, one could likely set this value at 10 percent or 0.10. Always make sure to set up the alternative hypothesis first.

For typical process improvement, a threshold P-value of 5 percent or 0.05 is used for Type I error. This threshold value is also called the standard of proof required or level of significance, denoted by the letter alpha (α). This standard is similar to proof beyond reasonable doubt. A lower level of significance will make it harder to prove one's point, and a higher level of significance will allow one to prove one's point more often. If the significance level is set at 0.05 or 5 percent, 5 percent of the time one will come to a significant difference by chance; 95 percent of the time the difference will be real.

The probability of making a Type II error is denoted by the letter beta (β). The main tool to use in reducing the chance of Type II error is to gather more data. Type II error consists of missing an improvement while some exists. Since the cost of missing an improvement is not as significant as the cost of Type I error (that is, the risk of saying there is an improvement when none exists), β is set at 10 percent, or 0.10.

The sample distribution is dependent on the null hypothesis. One assumes that this distribution is a normal distribution. For a large sample, this assumption is justified by the central limit theorem. Normal distribution is characterized, as usual, by its mean and its standard deviation.

After computing a value for t-statistics (depending on sample size), one needs to determine a P-value or theoretical t-values. These values can be calculated either by using t-tables or by Microsoft Excel. The TDIST function in Excel does the P-value calculations for you. However, the TDIST function does not allow you to enter negative values. For example, if you wish to find the P-value corresponding to -1.7, or your estimator is 1.7 standard deviations below the value in the null hypothesis, you are not able to enter -1.7. Since t-distribution is symmetric, the area above -1.7 is the same as the area below 1.7, which equals 1-TDIST (1.7, degree of freedom, 1). In this case, the test statistics have an approximate t-distribution with a mean of 0 and $n - 1$ degrees of freedom.

So far we have used the *t*-distribution. However, the normal distribution is used for testing the hypothesis for population proportions. You follow the same procedure to formulate test statistics for population proportions, as mentioned before. However, you use a different formula. Population proportion will follow the normal distribution, not a *t*-distribution. The Excel function used in this case is NORMSDIST.

STEPS FOR TESTING A HYPOTHESIS

Step 1. *Set up two hypotheses.* The theory of hypothesis testing is simply a statistical formalization of commonsense procedures. It makes the following assumptions:

1. Make a statement about the desired change H_a.
2. Make a statement of no change H_0.

Types of Hypotheses

One-sided, greater than hypothesis:

$$H_0: \mu \leq 5 \qquad H_a: \mu > 5$$

One-sided, less than hypothesis:

$$H_0: \mu \leq 5 \qquad H_a: \mu < 5$$

Two-sided, not equal to hypothesis

$$H_0: \mu = 5 \qquad H_a: \neq 5$$

The type of hypothesis will determine the type of test, i.e., the right tail, the left tail, or the two tailed. When the two-tailed test is used, the α-risk (e.g., 0.05) is distributed on both sides of the distribution.

Errors

Type I.   Rejecting null hypothesis when null is true
Type II.   Failing to reject null hypothesis when null is false.

Regardless of the population parameters, hypothesis testing includes the following steps:

1. Identify what you would like to prove and make that statement your alternative hypothesis.
2. Set up the null hypothesis.
3. Determine the level of significance α.
4. Gather data and calculate the t- or Z-value for sample distribution.
5. Determine the P-value using the appropriate function, TDIST, or NORMSDIST.
6. Compare the statistic to the theoretical values, or the P-value to the level of significance.
7. Draw a conclusion. Reject the null hypothesis if the P-value is smaller than the significance level; otherwise, null is not rejected.

To evaluate the data, begin by assuming that the null hypothesis is correct. The null hypothesis determines the sampling distribution of the estimator. Assume that this distribution is a normal distribution and that the assumption is justified by the central limit theorem. Conducting a one-tailed or two-tailed test, the procedures are as follows:

One-Tailed Test.   When the null hypothesis is an inequality, the hypothesis test is called one-tailed (left or right tail).

Alternative hypothesis.   Average number of defects will exceed 30.
Null hypothesis.   Average number of defects will be less than or equal to 30.

If you are able to reject the null hypothesis with an equivalent value of 30, the test will give you a lot of confidence in your conclusions. Generally, the equal sign is associated with the null hypothesis and the value used for the t- test.

Step 2.   *Calculate t-statistic.* Here is the formula to calculate the t-statistic:

$$t = \frac{\text{estimator} - \text{equal value in null hypothesis}}{\text{standard deviation of estimator}}$$

In essence, t measures how far the observed estimator is from the value one would expect if the null hypothesis were true. This distance is measured in terms of standard deviation (similar to the Z-value concept). The interpretation of t could therefore be expressed as follows: The estimate is the value of the statistic standard deviation away from the value in the null hypothesis. Test statistics will have a t-distribution.

Step 3.   *Calculate P-value.* Find the P-value using Excel or t-tables. Use the TDIST function if you have t-distribution. The Excel TDIST function screen is shown in Fig. 6.16. Use the t-statistic value for x. The degree of freedom is equal to $n - 1$ and this is a one-tailed test, so tails $= 1$.

Step 4.   *Compare P-value with the level of significance.*

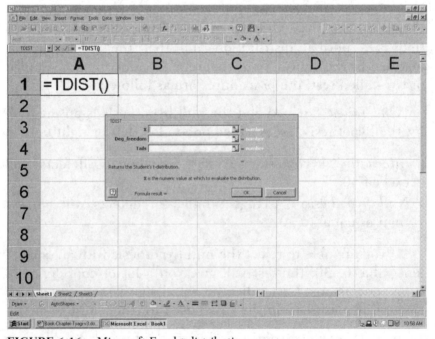

FIGURE 6.16.   Microsoft Excel t-distribution screen.

P-*value is lower than the level of significance.*   Accept the alternative hypothesis.

P-*value is higher than the level of significance.*   Do not accept the alternative hypothesis, or choose the null hypothesis.

Two-Tailed Test.   Suppose you want to prove that sales for your product are not equal to 30 units.

Alternative hypothesis.   Sales are not equal to 30 units.

Null hypothesis.   Sales are equal to 30 units.

This test is a bit different from the previous test. The null hypothesis is not an equation. This type of test is called a two-tailed test. The procedure for calculating the P-value is a bit different for a two-tailed test. The last Excel TDIST parameter, describing the number of tails, equals 2 for a two-tailed test.

One-tailed test:

$$TDIST (.., ..., 1)$$

Two-tailed test:

$$TDIST (..., ..., 2)$$

As shown in Fig. 6.17, two-tailed test statistics and P-value calculations include areas on both sides of the probability curve.

As an example, someone claims that the weight of a pack of pellets is 5 kg or less. Take the following steps to test this claim:

Step 1.   *Set up two hypotheses.* The alternative hypothesis is that the weight of a pack of pellets is less than 5 kg, and the null hypothesis is that the weight of a pack of pellets is 5 kg or more. Two hypotheses are written as follows:

$$H_0: \mu \geq 5$$
$$H_a: \mu < 5$$

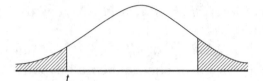

FIGURE 6.17.   Two-tailed *t*-test.

Step 2. *Calculate t-statistics.* You need the sample mean and standard deviation of the estimator or the standard error of the mean—or, in other words, the standard deviation of the sample. Assume that the sample has 30 observations, sample mean = 6, and standard deviation of the estimator = 0.5. Use the following formula to calculate the *t*-statistic:

$$t = \frac{\text{estimator} - \text{equal value in null hypothesis}}{\text{standard deviation of estimator}} = \frac{6 - 5}{0.5} = 2$$

Step 3. *Calculate P-value.*

$$t = 2.0$$

Degree of freedom = 30 − 1 = 29

$$\text{Tails} = 1$$

$$P = \text{TDIST}(2.0, 29,1) = 0.027$$

Step 4. *Compare P-value with the level of significance.* Assume the value of the level of significance is 0.05 or 5 percent. The *P*-value is lower than the level of significance, so the null hypothesis is rejected. The *P*-value area calculated in Step 3 is shown in Fig. 6.18.

   In another example, we consider population proportions. We want to compare two populations and find out whether there is a significant difference in terms of the proportions of defective parts.

   Defective population proportion for sample 1 = p_1
   Defective population proportion for sample 2 = p_2

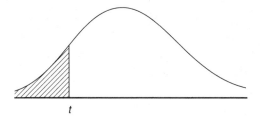

FIGURE 6.18.   *P*-value probability area.

Step 1.   *Set up two hypothesis.* We want to prove that the defective proportion differs between samples 1 and 2. The alternative hypothesis is that the difference between the defective proportions is not 0. The hypotheses:

$$H_0: p_1 - p_2 = 0$$
$$H_a: p_1 - p_2 \neq 0$$

Step 2.   *Calculate* t-*statistic or Z-value.* The procedure for calculating the *t*-statistic is similar, with the exception of the standard deviation calculation. For the sample proportions, we have a different formula to calculate standard deviation:

$$S_{\overline{p}_1 - \overline{p}_2} = \sqrt{\frac{\overline{p}_1(1 - \overline{p}_1)}{n_1} + \frac{\overline{p}_2(1 - \overline{p}_2)}{n_2}}$$

The test statistic is:

$$z = \frac{\overline{p}_1 - \overline{p}_2 - (p_1 - p_2)}{S_{\overline{p}_1 - \overline{p}_2}}$$

The distribution is normal; therefore we calculate a Z-value rather than a *t*-statistic.

Step 3.   *Calculate* P-*value.* Since the distribution is normal, we use the Excel NORMSDIST function to calculate the *P*-value in this case. The Excel function for normal distribution is shown in Fig. 6.19.

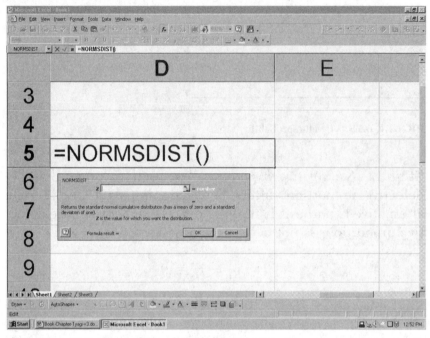

FIGURE 6.19.   Microsoft Excel normal distribution screen.

Assume that we end up with a *P*-value of 0.0456.

Step 4.   *Compare* P-*value with the level of significance.* We can reject the null hypothesis at a 5 percent level of significance (since $P < 0.05$); that is, we can be 95 percent confident (but not much more) that there really is a difference.

EVALUATING MEANS: ANALYSIS OF VARIANCE

The analysis of variance (ANOVA) test was invented by Ronald A. Fisher in 1920. The test compares two sets of data and can guide analysts in determining whether a certain event is most likely due to the random chance of natural variation or otherwise. The *F* ratio, named after Fisher, is the probability information produced by this test. Even though you are comparing means, you are in fact analyzing or comparing variances, therefore the name. Similar tests could be performed for two populations. The overall idea is similar to paired-sample t-testing.

Why not use a t-test in place of ANOVA? The answer lies in the number of pairs you need to test. For example, for 7 groups, you will have 21 pairs. ANOVA puts all the data into one number, called the F-test, and you don't have to deal with the complexities of multiple paired t-tests. On the other hand, unlike regression, ANOVA does not assume linear relationships and handles interaction effects automatically.

Test hypotheses for ANOVA could be written as follows:

$$H_0: \mu_1 = \mu_2 = \mu_3 = \cdots = \mu_k$$
$$H_a: \text{at least one of the } \mu\text{s is different}$$

This hypothesis means that for independent populations, samples are obtained from k levels of a single factor for the purpose of testing. The purpose of the hypothesis testing is to identify whether the k levels have equal means. A *factor* is defined as a quantity under examination in an experiment as a possible cause of variation in the response variable. Levels are the categories, measurements, or strata of a factor of interest in the current experiment. An ANOVA test with only one factor is termed a one-way ANOVA or one-factor ANOVA.

The total sum of squares (TSS) can be partitioned into two components—the sum of squares between samples (SSB) and the sum of squares within samples (SSW)

$$TSS = SSB + SSW$$

The ANOVA test depends on the following three parameters:

1. Size of the difference between group means.
2. Sample size in each group. Generally speaking, larger samples will tend to give more reliable data.
3. The variance of the dependent variable.

One-way ANOVA or univariate ANOVA includes single-classification ANOVA, or one-factor ANOVA. The design deals with one independent variable and one dependent variable.

Hypothesis testing can determine whether two populations have the same mean when the samples from the two populations are independent. We often encounter situations in which we are interested in determining whether three or more populations have equal means. A *t*-test will still be useful, and the tool gets very complicated. ANOVA is required for this type of hypothesis testing.

There are many different ANOVA designs to fit different situations. One-way ANOVA is the simplest.

Conditions for Applicability of ANOVA Methods

1. You must be able to regard the groups of observations as random samples from their respective populations.
2. The population variances must be equal.
3. The *k* samples must be independent of each other.
4. The observations within each sample must be independent of each other.
5. The *k* populations must be (approximately) normal with equal SDs.
6. Condition 4 is less crucial if the sample sizes n_i are large and are approximately equal.

Steps in ANOVA Analysis

1. Enter data appropriately and select the appropriate data set.
2. Use appropriate software to perform the calculation.
3. Select data and calculate the *F*-value.
4. Compare the *F*-value with the critical *F*-value.

Assumptions of normality and equal variance can be tested by developing histograms for the sample data from each population. The histogram should be bell-shaped, and the spread should be about the same for each population.

The *F*-test becomes less reliable as sample sizes become smaller, group sample sizes become more divergent, and the number of factors increases.

ANOVA TESTING

The variables that are measured are called *dependent* variables. The variables that are manipulated or controlled are called *factors* or *independent* variables. The test is based on the following three assumptions:

1. All populations are normally distributed.
2. The population variances are equal.
3. The observations are independent.

Imagine a machine that produces a steel pipe of a fixed diameter. You suspect that, depending on the mechanic, there could be variations in the diameter of the pipe. There are seven mechanics at work. You collect data from each mechanic and keep them in seven separate groups. Each group might have different highs and lows, and the averages might be different. But is this difference real?

ANOVA tests this difference. The comparison between the actual variation of the group averages and that expected from the formula is expressed in terms of the *F*-ratio, defined as follows:

$$F = \frac{\text{found variation of group averages}}{\text{expected variation of group averages}}$$

Test hypotheses for ANOVA could be written as follows:

$$H_0: \mu_1 = \mu_2 = \mu_3 = \cdots = \mu_k$$

H_a: at least one of the μs is different

Thus, if the null hypothesis is correct, one would expect F to be about 1, whereas a "large" F indicates the effect of variation due to a particular mechanic. How big should F be before you reject the null hypothesis? You need to have a threshold significance level. The table for a one-way ANOVA test is depicted in Fig. 6.20.

$$SS = \text{sum of squares of the variation}$$

$$DF = \text{degree of freedom}$$

$$MS = \text{mean squares of the variance} = \frac{SS}{DF}$$

$$F = \frac{MS}{MS\ (\text{within})}$$

Assume that the critical F-value based on a threshold value is 4. The computed value of F is 10, which is considerably more than 4.

As an example, consider a test to determine whether four different brake systems have different mean stopping distances.

Hypotheses for test:

$$H_0: \mu_1 = \mu_2 = \mu_3 = \mu_4$$

H_a: at least two population means are different

You could check the hypothesis at $\alpha = 0.05$.

From the hypothesis, recall that the decision rule is as follows:

One-Way ANOVA Table				
SOURCE OF VARIATION	SS	DF	MS	F
Between or explained				
Within				10.0
Total				

FIGURE 6.20.   One-way ANOVA test.

If $P < \alpha$, reject null hypothesis.

Otherwise, do not reject null hypothesis.

Note from Fig. 6.21 that the calculated F-value of 3.88 is greater than the F-critical value of 2.86; hence, you reject the null hypothesis and conclude that the means are not all equal. Figure 6.21 provides all the information that you need about the F-test. It shows the P-value associated with this test in the P-value column. You use the same logic as in hypothesis testing to reject or accept the null hypothesis.

Very often, in the absence of an F-critical value, you will have only F-statistics given, similar to t-statistics, and will have to calculate the corresponding P-value. In this case, use the FDIST function in EXCEL. Click Insert → Function, then choose FDIST as the function name. Enter the value of F next to X, enter $p - q$ (i.e., the number of variables being tested) next to Deg_freedom1 and enter $n - p - 1$ next to Deg_freedom2. As the Formula result Excel, will give the P-value.

p = independent variables in the extended model

q = number of variables in the base model;

thus $p - q$ = number of variables being tested

n = number of observations

$P = \text{FDIST}(X, p - q, n - p - 1)$

Alternatively, the F-value can be compared with the critical F-value corresponding to the significance level.

Source of Variation	SS	DF	MS	F	P-Value	F-Critical
Between groups	699.2	3	233.1	3.88	0.0166	2.86
Within groups	2,159.4	36	59.9			
Total	2,858.6	39				

FIGURE 6.21. ANOVA test results for brake system problem.

ANOVA Using Excel

First, enter data into the appropriate cells. Next, select "Anova: single factor" from the analysis tools and click OK. Select "Input variable range." After making the appropriate selection, the output table shows the source of variance as "Between groups" (= between treatments) and "Within groups" (= residual). The value of F is also provided. The F-value needs to exceed F-critical in order to have a significant difference between treatments.

BREAKTHROUGH SOLUTIONS— IMPROVE PHASE

After analyzing and understanding the sources of variation, the Improve phase enables actions to reduce variation or solve the problem. The Improve phase offers the opportunity to challenge the given, to question the status quo, and to look into some breakthrough solutions through idea generation and experimentation. Many nonstatistical or statistical tools can be used in the Improve phase. In order to achieve breakthrough solutions, the following questions must be answered:

- How does one gather the improvement ideas?
- What is the theory of innovation?
- How does one conduct experiments?
- How are the experimental results analyzed?
- What is a robust design?
- What are the factors for causing variation?

The tools most often used in the Improve phase include TRIZ for innovation, design of experiments to identify causes, comparative experiments to validate process change, comparative F- and t-tests, and Taguchi's loss function and response surface methodology for robust design and optimization.

OVERVIEW

Shrinking margins due to increased competition are forcing all organizations to become innovative and creative in providing value to customers. The essence of success lies in the speed of innovation and creativity. This requires full utilization of the total available mental power (TAMP) of all the stakeholders. This section discusses two main themes, *idea sourcing* and *innovation*.

IDEA SOURCING

The ideas may flow in from anywhere; there is a need to be sensitive to their reception and the related action. All stakeholders of the organization—the customers, employees, suppliers, competitors, owners, regulators, and community—have been regular contributors of ideas, and these ideas have helped the organization grow. However, the tools of the past may not necessarily be the tools for the future. In the past, maybe some ideas were ignored or missed for exploitation, but now, with the increased competition due to globalization, no one can afford such lapses. Now is the time that every single opportunity needs to be reviewed with maximum seriousness.

You need to capture the ideas of all stakeholders in a systematic manner. Following are some of the opportunities related to stakeholders, which may be low-hanging fruit waiting to be exploited.

Customers. Customers are the most direct source of new ideas related to existing and new products and services. Customer surveys may not always reflect a true picture, because the user and the person filling out the survey form are not always the same. All customer complaints are, however, potential opportunities. The normal tendency in any organization is to avoid the customers who complain a lot, but in reality, they are the organization's true friends, and all effort must be made to understand the problems and issues beyond the complaints.

Meeting with customers, postmeeting analysis of the opportunities, and building further on the analysis is an important method for generating ideas. It involves thoroughly understanding the customer's needs and working with the customer at all levels. There have been instances when operators at the customer's premises have provided excellent ideas, resulting in significant success.

Employees.   Employees have the potential to provide ideas on everything the organization does to provide value to its customers. The ideas may range from simplifying the logistics of serving the customers to complete redesigning of the products and systems to provide value in the most economical fashion. Employees are the most enlightened about waste in processes, most of which may be hidden from management. After all, the employees are working all the time in different functional areas to provide value to the customers. They are in the best position to answer the following questions:

- What are they doing?
- Where are the major loopholes?
- Which process is the bottleneck?
- What needs to be done to improve it?
- What are the additional opportunities?

Employees, by virtue of their experience, also have the best mental database, which may not be very structured but is of immediate utility for problem solving with less processing effort. The only need is to convert their experience to an actionable asset by harnessing their mental power.

Suppliers.   Due to globalization, falling trade barriers, and the resulting increase in competition, the trend is increasingly toward outsourcing and offshoring. While this enhances the importance of suppliers, it also signals the opportunities that lie with the suppliers.

Suppliers are the experts of their own businesses, and by virtue of their interaction with multiple clients, they have the knowledge pool regarding practices and many success stories to share. Even with compliance to the necessary confidentiality agreements, there are opportunities for the suppliers to share certain best practices.

A lot has been said about supplier management in terms of value engineering and refining the supply chain to reduce the final cost of the inputs. There is a need to focus on the softer issues of the customer–supplier relationship to build confidence. This would result in sharing the best practices of the industry, keeping the organization abreast of the latest trends and new opportunities. It can be achieved by refining the supplier performance evaluation system, which emphasizes the importance of providing new ideas to the organization to improve its competitiveness, motivating suppliers to share the success resulting from their efforts.

It may be a good idea to motivate high-performing suppliers by rewarding them with more business and eliminating the poor-performing suppliers. The reduction of suppliers may be more drastic, so that there is room to add a few new high-performing suppliers who can become potential success partners.

Competitors. Competitors are the most valuable asset any company can ever own. While on one hand they frustrate the company in the marketplace by snatching the orders, on the other hand they become the biggest force for innovation and progress. The ideal state for any organization would be to become its own best competitor—present versus past.

The very existence of competitors forces everyone in the organization to think and act differently. The degree of creativity is directly proportional to the amount of competitiveness in the industry. Observing and studying the competitor's actions and products may give an organization many good ideas and may lead to innovation and creativity.

While it is important to recognize the direct competition, it is equally important to recognize the indirect and future competition. As an example, consider metal drum manufacturing. There is no doubt that the other metal drum manufacturers are immediate competitors, but every manufacturer's business may be at risk if the customers switch to plastic drums. Here the competition comes from substitute products.

Owners. Owners are the individuals and institutions that have invested their money in the organization for faster growth than the rate of inflation. They naturally have the right to expect the desired returns. This puts things in a different perspective:

- They have the right to expect and demand success.
- They have the choice to put their money in alternate opportunities if their expectations are not met.

In order to avoid the flight of capital into alternate opportunities, again, the organization is forced to meet the expected demand—not only today, but every day. This also acts as a big force for the new idea generation.

Regulators. Traditionally, regulators have been viewed only as those who demand compliance with multiple laws and regulations. In reality, there are a number of opportunities for working with the regulators. While many of them may become the organization's actual customers, there may be a chance to get information from them regarding additional opportunities, both local and overseas.

There are a number of websites, such as www.cia.gov and www.nist.gov, that are full of information and tools that can be used in studying and exploiting the opportunities.

Community. The community, in general, places a lot of expectations on the organization. When interacting with community leaders, there are opportunities for understanding their further needs and fulfilling them more efficiently. This may become an additional source of opportunity and ideas for the organization.

INNOVATION

As the name suggests, innovation theory is a systematic approach that increases the probability of success. This probability is much higher than that provided by the traditional trial-and-error method. TRIZ is the most powerful tool used by inventors.

What is TRIZ?   TRIZ (theoria risheneyva isobretatelshehuh zadach) is a Russian acronym for *theory of inventive problem solving*. It is a tool that provides a dramatic breakthrough in almost all areas and accelerates the project team's ability to solve problems.

Background.   TRIZ was invented and structured by Genrich Altshuller, a patent examiner for the Russian navy. He recognized, after seeing hundreds of thousands of patent disclosures come across his desk, that there were a relatively small number of problem-solving principles that were repeatedly reused across many different areas of science and technology. When Altshuller completed his research of the world patent base, he had identified four key learnings:

Five Levels of Innovation.   Altshuller defined five levels of creativity based upon the degree of innovation. The lowest degree of innovation, such as a simple improvement, was assigned Level 1, and the discovery of a new phenomenon was assigned Level 5. A brief description of the various levels is presented in Fig. 7.1.

Innovations involving Levels 1, 2, and 3 are usually transferable from one discipline to another. This means that most of the inventive problems in any particular field have already been solved in some other field. The majority of patents fall within four major technologies: mechanical, electromagnetic, chemical, and thermodynamic.

Two Categories of Contradiction.   Inventive problems contain at least one contradiction. TRIZ recognizes two categories of contradictions:

- *Technical contradictions.*   These occur when you are trying to improve one characteristic of a technical system and simultaneously cause another characteristic to deteriorate. The desired state can't be reached because something else in the system prevents it. For example, you need a strong (desirable characteristic) material to hold, but it would become heavy (undesirable characteristic). Normally, a compromise

LEVEL	BRIEF DESCRIPTION
1	A simple improvement over the present based on the knowledge existing within the organization.
2	A significant improvement based on the knowledge existing within the industry.
3	A major improvement based on the knowledge across industries.
4	A new technology exploiting the knowledge from different fields of science.
5	Discovery of a new phenomenon exploiting knowledge from the universe.

FIGURE 7.1.   Levels of innovation.

solution is considered, but this is against the principle of TRIZ.

Note from Fig. 7.2 that TRIZ tends to move the barrier closer to the ideal state. TRIZ research has identified 40 principles, discussed later, to solve the technical contradictions.

• *Physical contradictions.* These occur when two opposite properties are required from the same element of a technical system or from the technical system itself. The four principles of separation, discussed later, solve the physical contradictions.

Standard Patterns of Evolution. TRIZ provides the first understanding of the trends, or patterns, of the evolution of technical systems. It is now an international science of creativity that relies on the study of the patterns of problems and solutions, not on the spontaneous creativity of individuals or groups. Millions of patents have been analyzed to discover the patterns that predict breakthrough solutions to technical problems.

The basic premise of TRIZ is that there is less difference among many areas of science and technology than are generally perceived and that problem solving, in a very general way, can be transferred from one area of science and technology to another. The basic premise of TRIZ is presented in Fig. 7.3.

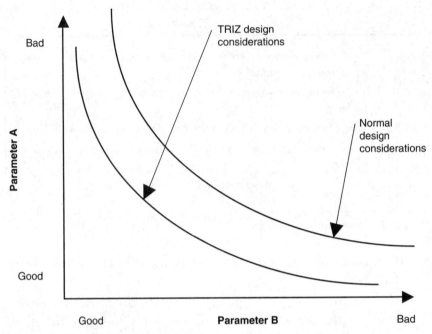

FIGURE 7.2.   TRIZ and normal design considerations.

TRIZ research began with the hypothesis that there are universal principles of invention that are the basis for creative innovations that advance technology, and that if these principles could be identified and codified, they could be taught to people to make the process of invention more predictable. The research has proceeded in several stages over the past 50 years. The three primary findings of this research are as follows:

1. Problems and solutions repeat across industries and sciences.
2. Patterns of technical evolution repeat across industries and sciences.
3. Innovations use scientific effects outside the field where they were developed.

The same principles are used in many inventive designs.

There are solution patterns, and the same fundamental solutions are used over and over again, and these implementa-

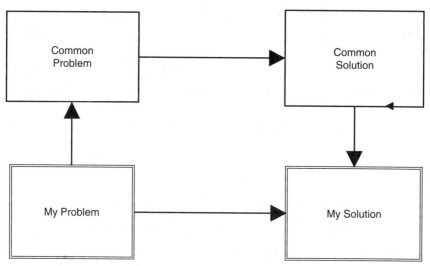

FIGURE 7.3.   Basic premise of TRIZ.

tions are often separated by many years. If some means of accessing the applications of these fundamental solutions were made available to inventors, the number of years between applications would decrease. Thus, the innovation process can become more efficient, the time between improvements is decreased, and the gap between various disciplines is narrowed.

TRIZ Methodology

Define the Problem in the Most Generic Terms.   It is important to define the problem correctly by stating the desired outcome, understanding all its elements and the desirable improvements. It needs to be defined in the simplest possible form. This helps to locate solutions that may already be present somewhere in the system.

Define the Ideal State.   The ideal design is a goal. The law of ideality states that any technical system, through its lifetime, tends to become more reliable, simple and effective—more ideal. Each improvement brings the system closer to its ideal performance; that is, it performs more tasks, costs less, decreases in size, wastes less resources, and so on.

The TRIZ approach to a problem leads one to think of performing a new function without introducing a new complexity to the system. The very first question asked in a TRIZ problem-solving session is "What is the ideal system?" The true definition would be something that performs its function without actually existing. The true success of inventing would be the removal of barriers to ideality; various routes to doing so are the following:

- Increase the functionality of the system.
- Transfer maximum action at the system output stage.
- Transfer some functions up or down the value chain.
- Use preexisting resources.

Identify the Contradictions. The contradictions were already mentioned in an earlier section. You need to identify both types, technical and physical.

Look for contradictions in the contradiction matrix. A sample of the matrix is depicted in Fig. 7.4. Find the row that most closely matches the feature or parameter you are improving in your trade-off, and the column that most closely matches the feature or parameter that degrades. The cell at the intersection of that row and column will have several numbers. These are the identifying numbers for the principles of invention that TRIZ research indicates are most likely to solve the problem; that is, to lead to a *breakthrough* solution instead of a trade-off.

Use the 40 Inventive Principles. In reviewing the thousands of patents, Altshuller distinguished between incremental or "routine" inventions and true breakthrough inventions. These were the inventions he reviewed to establish the 40 inventive principles. These breakthrough inventions invariably resolved a significant operational or design contradiction. What we normally do with contradictions is to compromise. Design or operational characteristics that allow a contradiction to be resolved truly constitute a breakthrough. These principles may broadly be classified as follows:

Undesirable Result → / Desirable Result ↓	1 Weight of moving object	2 Weight of non-moving object	3 Length of moving object	4 Length of non-moving object	5 Area of moving object	6 Area of non-moving object	7 Volume of moving object	8 Volume of non-moving object
1 Weight of moving object			15, 8, 29, 34		29, 17, 38, 34		29, 2, 40, 28	
2 Weight of non-moving object				10, 1, 29, 35		35, 30, 13, 2		5, 35, 14, 2
3 Length of moving object	15, 8, 29, 34				15, 17, 4		17, 7, 4, 35	
4 Length of non-moving object		35, 28, 29, 40				17, 7, 10, 40		35, 8, 2, 14
5 Area of moving object	2, 17, 29, 4		14, 15, 18, 4				7, 14, 17, 4	
6 Area of non-moving object		30, 2, 14, 18		26, 7, 9, 39				
7 Volume of moving object	2, 26, 29, 40		1, 7, 4, 35		1, 7, 4, 35			
8 Volume of non-moving object		35, 10, 19, 14	19, 14	35, 8, 2, 14				

FIGURE 7.4.   Sample contradiction matrix.

- *Principles involving actions on the parts.* Here appropriate action may be analyzed by focusing on the part. The examples include the following:

  Principle 1. Segmentation. Divide an object into independent parts; make an object easy to disassemble; increase the degree of fragmentation or segmentation.

  Principle 2. Taking out. Separate an interfering part or property from an object, or single out the only necessary part (or property) of an object.

  Principle 5. Merging. Bring closer together (or merge) identical or similar objects; assemble identical or similar parts to perform parallel operations, make operations contiguous or parallel; bring them together in time.

- *Principles involving the analysis of the environment.* In these principles, the focus is shifted from the part to the environment under which the part operates. The examples of this type include:

Principle 12.   Equipotentiality.   In a potential field, limit position changes (e.g., change operating conditions to eliminate the need to raise or lower objects in a gravity field).

- *Principles invoking counterintuitive thinking.*   In these principles, the thought process is provoked in unconventional manner. The examples of this type include:

  Principle 8.   Antiweight.   To compensate for the weight of an object, merge it with other objects that provide lift; to compensate for the weight of an object, make it interact with the environment (e.g., use aerodynamic, hydrodynamic, and other forces).

  Principle 9.   Preliminary antiaction.   If it will be necessary to do an action with both harmful and useful effects, this action should be replaced with antiactions to control harmful effects; create beforehand stresses in an object that will oppose known undesirable working stresses later on.

  Principle 13.   The other way round.   Invert the actions used to solve the problem (e.g., instead of cooling an object, heat it); make movable parts (or the external environment) fixed, and fixed parts movable; turn the object (or process) upside down.

- *Principles involving the mechanical properties of the system.* Here the focus is on the mechanical properties of the system. The examples include the following:

  Principle 18.   Mechanical vibration.   Cause an object to oscillate or vibrate; increase its frequency (even up to the ultrasonic); use an object's resonant frequency; use piezoelectric vibrators instead of mechanical ones; use combined ultrasonic and electromagnetic field oscillations.

  Principle 28.   Mechanics substitution.   Replace a mechanical means with a sensory (optical, acoustic, taste, or smell) means; use electric, magnetic, and electromagnetic fields to interact with the object; change from static to movable fields, from unstructured fields to those having structure; use fields in conjunction with field-activated (e.g., ferromagnetic) particles.

Principle 29. Pneumatics and hydraulics. Use gas and liquid parts of an object instead of solid parts (e.g., inflatable, filled with liquids, air cushioned, hydrostatic, hydroreactive).

Use the Four Separation Principles. TRIZ has four classical ways to resolve physical contradictions:

- *Separation in time.* When a plane is landing or taking off, the landing gear must be present, but the gear must be absent when the plane is cruising to reduce the air drag. This means that the landing gear must be present at one time and absent at another. The problem has been solved by separating the requirements in time and making landing gear retractable.

- *Separation in space.* In a coating process that entails dipping parts in a bath of coating solution, the output can be increased by increasing the solution temperature. Doing so, however, reduces the usable lifetime of the solution bath. The principle of separation in space suggests that the bath solution be cold but that the part be hot. Here the requirement can be separated in space—hot part and cold bath—and the desired result may be achieved by dipping the hot part in the cold bath.

- *Separation based on condition.* One example is to change the physical state of the matter. For example, in order to fill candy with liqueur, the syrup is frozen and dipped in the candy syrup. Another example is to change the magnetic properties of the matter. For example, a process uses Curie temperature to change the properties from paramagnetic to ferromagnetic in order to sensitize the system to the temperature.

- *Separation between the whole system and its parts.* This concept may be illustrated with a nontechnical example. An organization needs to be large enough to have an economy of scale for its operations. But large companies tend to become more bureaucratic, hence retarding the quick responses needed in today's world. So the organization needs to be big and small. This is a contradiction. To solve this contradiction, this principle may be used. Form strategic business units within an overall umbrella of the corporate office. The corporate office would have minimal interference in routine matters and would provide only policy guidelines to the units.

COMPARATIVE EXPERIMENTS— CURRENT VERSUS BETTER

You need to test the results of the experiments to confirm whether they have significantly improved the desired output from the present state. You normally come across two kinds of outputs for evaluation:

1. The output follows a particular distribution (i.e., parametric).
2. The output does not follow a particular distribution (i.e., nonparametric).

   This section examines the tools available to evaluate the improvement of both types.

PARAMETRIC OUTPUT

The following assumptions are made about the output data under evaluation:

1. The two sets of samples, as a result of the current process parameters, and improved process parameters are independent and have been randomly selected from the population that is normally distributed.
2. The measurement error is negligible and will not affect the analysis.

   Before comparing the output data of the two samples, it is customary to compare their variances to ensure that they are not significantly different. This implies that when the means are compared, they can give an accurate estimate of the improvement sought by the experiment.

***F*-test.**   The standard procedure for comparing the variances is the *F*-test. To perform *F*-testing, we make the following hypotheses:

Null hypothesis H_0:

$$\sigma_m = \sigma_p$$

Alternate hypothesis H_1:

$$\sigma_m \neq \sigma_p$$

where σ_m = standard deviation of the samples obtained from the modified process conditions

σ_p = standard deviation of the samples obtained from the present process conditions

This implies that you need to perform a test that would disprove the equality of the two standard deviations. The statistic that is used to compare the variances is called the F-test statistic, defined as follows:

$$F = \frac{S_1^2}{S_2^2}$$

where S_1^2 = larger variance

S_2^2 = smaller variance

The calculated F-test statistic is compared with the critical F-value picked from the F-table. To pick the critical F-value, you need the following information:

1. Desired level of significance α. This establishes whether the difference is due to chance causes or due to assignable reasons. Since the test of hypothesis is at the two ends, pick the value from the table at $\alpha/2$.
2. Degree of freedom of the sample with the larger standard deviation $DF_1 = n_1 - 1$, where n_1 is the number of observations in the sample with larger standard deviation.
3. Degree of freedom of the sample with the smaller standard deviation $DF_2 = n_2 - 1$, where n_2 is the number of observations in the sample with smaller standard deviation.

This principle is illustrated by an experiment. Perform the test at the two process levels, current and modified as per the new process condition defined in Fig. 7.5. The process output is presented in Fig. 7.6.

Following are the values of standard deviation of the present and modified process outputs:

$$S_p = 6.4550$$

$$S_m = 16.5202$$

here S_1^2 = variance corresponding to S_m
S_2^2 = variance corresponding to S_p

That is

$$S_1^2 = 41.6667$$

$$S_2^2 = 272.9167$$

$$F = \frac{272.9167}{41.6667} = 6.5500$$

PROCESS PARAMETER	PRESENT	MODIFIED
Temperature, °F	90	100
Pressure, psi	80	90

FIGURE 7.5. Process parameters of the two samples.

PRESENT	MODIFIED
105	145
95	120
90	125
100	105

FIGURE 7.6. Process output.

Assume that you want to test the significance at 90. Find the critical F-value from the F-table that corresponds to $\alpha = 5$ percent, $DF_1 = 4 - 1 = 3$, and $DF_2 = 4 - 1 = 3$. This value is 9.2766. The critical F-value and the calculated F-value are presented in Fig. 7.7.

Since the calculated F-value is less than the critical F-value, you don't have enough evidence to reject the null hypothesis. Hence, you can conclude with 90 percent confidence that the standard deviations of the two samples are statistically the same, and proceed to the next step of confirming the improvements.

***t*-Testing the Difference of the Means.**   The *t*-test is performed to compare the following:

• Sample mean with the population mean
• Mean of the two samples

The basic assumptions in performing the *t*-test are as follows:

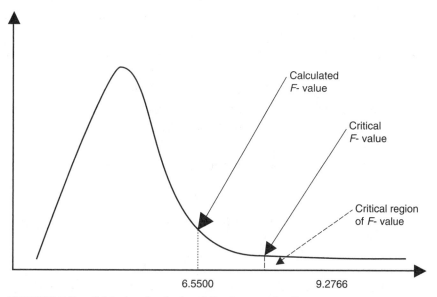

FIGURE 7.7.   Critical and calculated F-values on the F-curve.

1. The population distribution is approximately normal.
2. Samples were selected in a random fashion and are independent.
3. The measurement error is minimal.
4. The sample size is small (less than 30).
5. The population standard deviation is unknown.

When performing the t-test, make the following hypotheses:

Null hypothesis H_0:

$$\mu_m \leq \mu_p$$

Alternate hypothesis H_1:

$$\mu_m > \mu_p$$

where μ_m = mean value of the sample data of the modified process

μ_p = mean value of the sample data of the present process

This implies that you need to test whether the mean of the modified process is the same as or less than the present one. This is pictorially represented in Fig. 7.8.

In this experiment it is assumed that the improvement is in the positive direction. It may also happen that the improvement can be expressed in the negative direction. Examples of this type include the reduction in defects. In this case, the signs in the hypothesis change, as follows:

Null hypothesis H_0:

$$\mu_m \geq \mu_p$$

Alternate hypothesis H_1:

$$\mu_m < \mu_p$$

Sometimes the means need to be compared for equivalence. Examples include meeting the target value of the dimen-

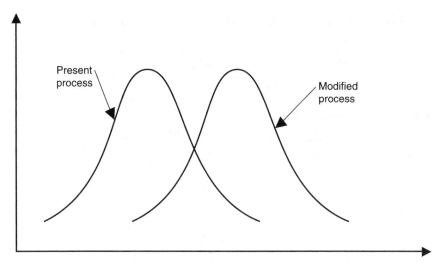

FIGURE 7.8.   Present and modified process.

sions by the two suppliers. In this case, the hypotheses would be as follows:

Null hypothesis H_0:

$$\mu_m \neq \mu_p$$

Alternate hypothesis H_1:

$$\mu_m = \mu_p$$

The generalized *t*-test statistic is calculated as follows:

$$t = \frac{\overline{X} - \mu}{s/\sqrt{n}}$$

where $\overline{X}$ = sample mean
μ = population mean
s = sample standard deviation
n = sample size

This formula is used when the population standard deviation is not known and you need to compare the sample with respect to the population.

The comparative *t*-test statistic is calculated as follows:

$$t = \frac{\overline{X}_m - \overline{X}_p}{S_{pl} \sqrt{(1/n_m) + (1/n_p)}}$$

where $\overline{X}_m$ = mean of the modified process
$\overline{X}_p$ = mean of the present process
S_{pl} = pooled standard deviation, calculated as follows:

$$S_{pl} = \sqrt{S_{pl}^2}$$

$$S_{pl}^2 = \frac{(n_m - 1) S_m^2 + (n_p - 1) S_p^2}{n_m + n_p - 2}$$

where S_m = standard deviation of the samples from the modified process
S_p = standard deviation of the samples from the present process

The preceding formula is applicable when the variances of the two populations are statistically the same. This is confirmed by performing the *F*-test, as discussed earlier.

The calculated *t*-statistic is compared to the critical *t*-statistic. The critical *t*-statistic is found as follows:

1. Decide the desired level of significance, α. This defines the probability of the desired difference due to assignable causes. These assignable causes are the modified process conditions.
2. Degree of freedom DF = $n_m + n_p - 2$

This example again refers to the process output data from Fig. 7.6. The sample means are as follows:

$$\overline{X}_m = 123.75$$

$$\overline{X}_p = 97.5$$

Here, since $n_m = n_p = 4$

$$S_{pl}^2 = \frac{S_m^2 + S_p^2}{2}$$

$$S_{pl}^2 = \frac{272.9167 + 41.6667}{2} = 157.2917$$

$$S_{pl} = 12.5416$$

Substituting these values in the formula of the calculated t-statistic:

$$t = \frac{123.75 - 97.5}{12.5416 \sqrt{(1/4) + (1/4)}} = 2.96$$

Now, compare this with the critical t-value. Assume that you want to test the improvement at a significance level of 95 percent. In this case, since the sample size is 4 in both the samples, the degree of freedom is $4 + 4 - 2 = 6$. The critical value of t, from the t-table, for $\alpha = 95$ percent and $DF = 6$, is 1.943. The critical t-value and the calculated t-value are depicted in Fig. 7.9. Note from this figure that the calculated value of t is in the critical region; this means that there is enough evidence to reject the null hypothesis with 95 percent confidence. This implies that the modified process is significantly better than the present process.

In the preceding example, the test is performed in one direction (increase of the output value), called a one-tailed test. In case you need to confirm the equality of the two process outputs, you need to ensure that the process output doesn't shift in either direction from the target. Thus, you need to perform the two-tailed test. When performing the two-tailed test, divide α into two equal parts and then locate the critical value corresponding to $\alpha/2$ from the t-table.

NONPARAMETRIC OUTPUT

There may be a situation in which you have very few samples of the output and don't know the nature of their distribution.

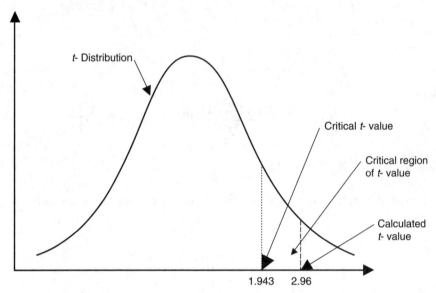

FIGURE 7.9.   Critical and calculated t-values.

Now you need to analyze the output differently. Under these circumstances, assume that the populations of the two sets (present and modified conditions) of samples have similar probability characteristics.

We shall illustrate this concept with an example. A school wants to make a quick evaluation of the effect of two different sets of textbooks (new is believed to be better) on the performance of two sets of students. Assume that the method of teaching, the level of students in each group, and all other conditions are the same. A comparison of their test scores over a quarter's time would confirm the validity of the improvement realized with the new set of textbooks. The monthly mean test scores of the students are listed in Fig. 7.10. Ranking the scores from most desirable to the least desirable, they appear as shown in Fig. 7.11. You may note that the new book does not show any appreciable improvement in student scores.

In experiments you come across two kinds of scenarios:

1. Every single observation of the output from the modified process is superior to the observation of any output from the present process (i.e., there is no overlap of the ranks).

SCORES WITH PRESENT BOOK P	SCORES WITH NEW BOOK M
90	92
85	89
91	88

FIGURE 7.10. Mean scores for the 3 months of the school quarter.

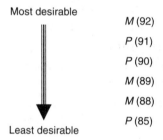

Most desirable

M (92)
P (91)
P (90)
M (89)
M (88)
P (85)

Least desirable

FIGURE 7.11. Ranking of scores.

2. The ranks of the observations overlap.

No Overlap of the Observed Data. You can imagine that for this scenario to be true, there must be a significant shift from the present level of performance. As for parametric tests, here also you formulate your hypothesis as follows:

Null hypothesis H_0. The modified process is the same as or worse than the present process.
Alternate hypothesis H_1. The Modified process is better than the present process.

Now you define certain terms.

$$\alpha \text{ risk} = \frac{1}{\text{total combinations}}$$

You divide one by the total number of combinations in calculating α risk because the improvement would be decided on the basis of only one combination out of all possible combinations of the experiments.

$$\text{Total possible combinations} = \frac{(n_m + n_p)!}{n_m!n_p!}$$

You know from the discussion of parametric output that n_m is the number of samples from the modified process and n_p is the number of samples from the present process.

The level of α risk (0.001 to 0.10) is chosen based on the seriousness of the implications of the incorrect decision. After arriving at the level of α risk, calculate the number of samples required from the modified and present scenarios to make the necessary decisions. The combination of the number of samples from the modified and present scenarios is decided based on the balance of the cost of producing the modified sample and the cost of performing the experiments.

After the number of samples from the modified and present process has been decided, the experiment is performed in a random order, the outputs are ranked, and then if all the modified outputs rank above the present ones, we say that the modification has passed the test.

Return to the example as we considered at the beginning of this section. Suppose that the school selects an α risk of 0.001 for the textbook change because the student's future is at stake. The combinations of the number of samples (in this case the average data of the number of months) will be as presented in Fig. 7.12.

Note from Fig. 7.12 that various combination options are available. The school needs to complete the project as soon as possible, and for that purpose six samples each is the optimum option, because the project can be completed in 6 months. Assume that the tests are performed and the output data are recorded as in Fig. 7.13.

The scores are ranked as shown in Fig. 7.14. Note that all modified process outputs are above the present process outputs. This implies that the new process is a significant improvement over the present process.

n_m	n_p	TOTAL
4	10	14
5	8	13
6	6	12
6	7	13
7	6	13
8	5	13
10	4	14

FIGURE 7.12.
Combinations for α risk
of 0.001.

SCORES WITH PRESENT BOOK P	SCORES WITH NEW BOOK M
90	92
85	93
91	92
90	91
91	94
90	93

FIGURE 7.13.   Mean scores for the 6 months of the project.

Overlap of the Observed Data.   It is very unlikely that in real life you will ever come across such a neat separation of the outputs of the modified and present processes as was shown in the scenario for no overlap of the observed data.

When you have an overlap, you have three regions:

1. Region of the modified end count
2. Region of overlap
3. Region of the present end count

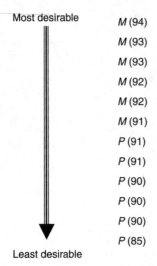

Most desirable

M (94)

M (93)

M (93)

M (92)

M (92)

M (91)

P (91)

P (91)

P (90)

P (90)

P (90)

P (85)

Least desirable

FIGURE 7.14.   Ranking of scores.

This is explained by a likely scenario in Fig. 7.15, which identifies these regions.

The decision rule in case of overlap is based only on regions 1 and 3. The region of overlap is ignored. In this case, the samples should be approximately equal in size, and the decision rule as shown in Fig. 7.16. As may be noted from this figure, like total number of end count (modified plus present process output) should be more than a critical value to claim the improvement. The purpose of this decision rule is to establish a clear concentration of the present and modified levels of the process output.

UNIVERSAL DESIGN OF EXPERIMENTS— COMPONENT SEARCH

In the previous section, you learned how to compare the improvement over the present process. Now in this section you will learn the generic rules for troubleshooting the problem and identifying the problematic parts of the system. This section discusses the process of identifying the troublesome parts in any assembly or subassembly. The key logic is a process of

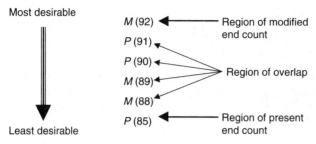

FIGURE 7.15.   The three regions of the end count.

α Risk	End Count
0.05	≥6
0.01	≥9
0.001	$12

FIGURE 7.16.   Decision rule for improvement.

elimination to identify the parts contributing to the inferior output.

The purpose of a component search experiment is to identify the components causing variation in the performance of an assembly. This type of experiment requires one *good* assembly and one *bad* assembly (the assembly that needs correction). The technique uses a process of elimination approach by switching the components between these two assemblies to determine the effect of each component on the output response being measured.

The goal of the experiment is to either establish that a component is causing the poor performance of the assembly or eliminate the component as a factor contributing to the performance problem.

KEY ASSUMPTIONS

1. There are two completed assemblies, one with all good parts and another with a few troublesome parts.

2. The difference in the performance of the two assemblies is significant and is solely caused by the difference in the performance of the components.

3. The performance is solely a function of the parts; i.e., the process of interchanging the parts does not affect the performance, and all the relevant parts are perfectly interchangeable.

4. Only those components that can be interchanged without any assembly/disassembly effects on performance can be investigated in this experiment.

5. The performance can be measured and repeated any number of times without deterioration.

PROCEDURE

1. Measure the performance output of the good and bad assemblies.

2. After the troublesome assemblies have been identified, brainstorm to identify the possible troublesome parts.

3. Rank all possible troublesome parts in the order of their effect on the assembly.

4. Interchange the most troublesome part between the good and bad assemblies.

5. Measure the performance of the output.

6. Analyze the results. There are three possible outcomes, which determine the next course of action:
   - *No change in the output.*   The interchanged part may not be the cause of poor performance. Interchange the components back to return to the old state. Now interchange the next ranked part between the assemblies, and continue with the experiment.
   - *Complete reversal in the output.*   The interchanged part was the root cause of the problem.
   - *Partial change in the output.*   This part is the cause of the problem but may not be the only one. There may be other parts that might also be causing the problem. Keep on

interchanging the parts in a combination that maximizes the reversal. The capping run ensures this.

CAPPING RUN

The objective of the exercise is to identify all the components that are adversely affecting the product or process performance. This ensures that all the troublesome elements are isolated for corrective action as the next logical step.

The capping run ensures that *all* the components that affect the difference in the product or process output have been identified. This is normally performed by completely interchanging all the parts that caused the complete reversal of the product or process performance.

Consider the example of two electronic assemblies—one good (denoted G) and the other bad (denoted B). Assume that you have measured the performance of the two assemblies. You know from the historical data that the most troublesome parts of this assembly could be the capacitor C, diode D, and resistor R, in order of their effect on the performance output. These components are identified as shown in Fig. 7.17. To identify the problem component, perform the following experiment.

First, interchange C_G and C_B and measure the difference in performance. You observe no difference in the performance output. This means that the capacitors are not causing the problem in this case, and you interchange the capacitors again to bring the assemblies back to their original state.

Interchange D_G and D_B and measure the performance difference. You observe a minor difference in the performance output. This means that the diodes are not the only factor causing the problem in this case. You interchange the diodes again to bring the assemblies back to their original state.

Interchange R_G and R_B and measure the performance difference. You observe some difference in the performance output, but not the complete reversal of the performance. This means that the resistors are a significant factor, but not the only factor causing the problem in this case. Now it will be interesting to see the interaction of the diodes and resistors on the performance output.

COMPONENTS	GOOD ASSEMBLY	BAD ASSEMBLY
Capacitor	C_G	C_B
Diode	D_G	D_B
Resistor	R_G	R_B

FIGURE 7.17.   Components of the two assemblies.

Now you interchange both D_G and R_G with D_B and R_B simultaneously, and measure the performance difference. You observe the complete reversal of the performance output. This means that both the diodes and the resistors were causing the problem and need to be addressed suitably.

DESIGN OF EXPERIMENTS—FULL FACTORIALS

Design of experiments (DOE) is a powerful statistical technique introduced by Ronald A. Fisher in England in the 1920s to study the effect of multiple variables simultaneously. In his early applications, Fisher wanted to find out how much rain, water, fertilizer, sunshine, and so forth are needed to produce the best crop. Since that time, much development of the technique has taken place in the academic environment, and this has generated many applications for the production floor.

The purpose of any process testing experiment is to choose an option to add more value, reduce the cost, or both. In the majority of the cases, the system, in general, is already working but needs the fine-tuning. Conventional experiments focus on one variable at a time and try to keep all other variables constant. This not only is unrealistic in complicated processes but also requires a large number of experiments to draw a conclusion. This also does not permit the exploration of the simultaneous effect of changing two variables. There is a possibility that when two variables are simultaneously changed in a certain way, the results may be dramatically improved.

These limitations are overcome in the DOE model, which permits changing more than one variable simultaneously, at multiple levels, and observing the combined effect on the outcome. The DOE model requires fewer experiments to draw a conclusion, which in turn saves time and expensive resources.

FULL-FACTORIAL EXPERIMENTS

The term *full factorial* means that all possible combinations need to be tested to draw an experimental conclusion. For the simplicity of understanding, assume that the experiments are conducted at two levels of N variables. The two levels may be regarded as high H and low L. Conducting the experiment at two levels gives a clear idea of the direction of the results without conducting a large number of experiments.

Why Two Levels?
1. The two levels help create a balance in the experiments, as the number of high levels and the number of low levels for each variable are the same.
2. The absolute level of the variable is also not very clear at the beginning of the experiment, and only their relationship is under review.

The number of experiments you need to conduct for the number of variables is given in Fig. 7.18. From this figure, we can make the possible combinations of the experiment as follows.

One Variable.   One variable X1 can be kept at two levels, low L and high H. You need to conduct experiments to determine the effect of only one variable at two levels, as shown in Fig. 7.19.

Two Variables.   The layout of the two variables is shown in Fig. 7.20.

Two variables, X_1 and X_2, can be varied at two levels, making four cells. Experiments are conducted at all the conditions

NUMBER OF VARIABLES	NUMBER OF EXPERIMENTS
1 X_1	$2^1 = 2$
2 X_1, X_2	$2^2 = 4$
3 X_1, X_2, X_3	$2^3 = 8$
4 X_1, X_2, X_3, X_4	$2^4 = 16$

FIGURE 7.18. Relationship between the number of variables and the number of experiments at two levels.

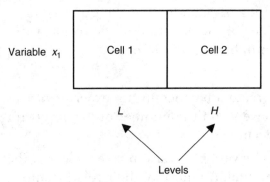

FIGURE 7.19. Combinations for one variable.

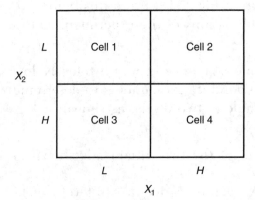

FIGURE 7.20. Combinations for two variables.

represented by these four cells with different combinations of the two variables.

Three Variables.   The layout of the three variables is shown in Fig. 7.21. The four corners of the box represent the eight conditions at which the experiments are performed.

More Than Three Variables.   The pictorial representation of the possible combinations of more than three variables is very complex, but you need to identify all the possible combinations to conduct a full-factorial experiment. Use the guideline for four variables in Fig. 7.22. The general pattern of H and L for N variables is shown in Fig. 7.23.

Note that as the number of variables increases, the level of complexity increases. At the same time, the large number of variables may not necessarily add to the degree of refinement of the conclusion. In real life the Pareto principle applies to all situations. This helps to isolate those variables that have a significant effect on the outcome of the experiment. In the majority of cases, there are two or three variables that have a significant effect on the outcome.

It takes experience to identify the correct variables, and this makes the experiment truly valuable. Normally, to minimize risk and achieve the purpose of experiment at the same time,

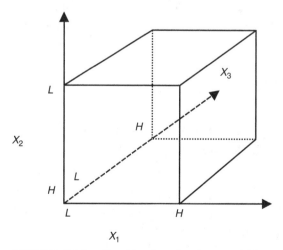

FIGURE 7.21.   Combinations for three variables.

Cell	X_1	X_2	X_3	X_4
1	L	L	L	L
2	L	L	L	H
3	L	L	H	L
4	L	L	H	H
5	L	H	L	L
6	L	H	L	H
7	L	H	H	L
8	L	H	H	H
9	H	L	L	L
10	H	L	L	H
11	H	L	H	L
12	H	L	H	H
13	H	H	L	L
14	H	H	L	H
15	H	H	H	L
16	H	H	H	H

FIGURE 7.22. Combinations for four variables.

the extreme values of the tolerance of the variables are taken for the initial experiment. This is further explained by the following example.

Steps for Conducting a Full-Factorial Experiment

A chemical process gives an output of 100 units per hour. Due to an increase in demand, management wants to increase the output without further investment.

Step 1. *Define the experiment objective.* The objective of the experiment is to establish the relationship of the process parameters to an increase in output.

Step 2. *Identify the key variables.* Process experts feel that the output is sensitive to the temperature and pressure at which

Column *N*	Column *N-1*		Column 1
X_1	X_2	....	X_N
L	L		L
L	—		H
—	(¼ of total combinations)		L
—	H		H
(½ of total combinations)	(¼ of total combinations)		—
H	L		—
H	—		—
—	(¼ of total combinations)		—
—	H		—
—	—		—
(½ of total combinations)	(¼ of total combinations)		—

FIGURE 7.23.   Combinations for N variables.

the reactions take place. This also modifies the objective of the experiment, which is now to study the effect of temperature and pressure in increasing the process output.

Step 3.   *Conduct a preliminary study of the expected response and decide on the levels.*

The purpose of this step is to identify the two levels of the experiment for both the variables. This establishes the response of the output with respect to the changes made in the levels of different variables and determines their levels for the purposes of the experiment.

This is one of the most crucial aspects of the experiment, particularly when the costs are high. The decision requires extensive experience and knowledge of the process. Usually, when no other alternative is available, it is suggested that you take the two extreme values of the tolerance of the process parameters, either as defined for the process or learned through experience. In the case of the experiment under discussion, the levels are given in Fig. 7.24.

Step 4.   *Conduct the experiment.*

Determine the Cells.   You need to conduct the full-factorial experiment to evaluate all the combinations of the extreme

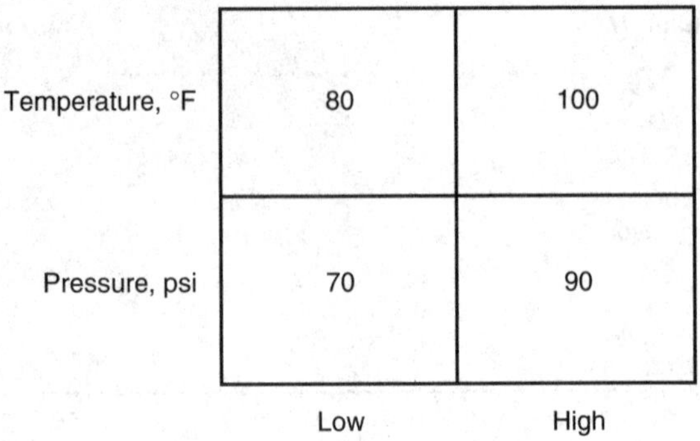

Temperature, °F	80	100
Pressure, psi	70	90
	Low	High

FIGURE 7.24. Two levels of the process parameters.

values of temperature and pressure. The possible combinations shown in Fig. 7.25 may be tried. This should not result in waste, as the experiment conditions are within the tolerance limits of the process parameters.

Note from the preceding example that by keeping the temperature constant at 80°F, you have two sets of observations in Cell 1 and Cell 2 for pressures of 70 and 90 psi. When pressure is kept constant at 70 psi, you get two sets of observations in Cell 1 and Cell 3 at 80 and 100°F. This means that you can use the observations for Cell 1 twice—that is, double the quantity of data are available for the purpose of this experiment, as compared to conventional experimentation. This power of the DOE method is called *hidden replication.*

Determine the Number of Samples per Cell. The number of samples required to test a hypothesis is a function of the following factors:

- Type I error
- Type II error
- Variance of the process
- Shift required to make the process significantly better

For α = 5 percent, β = 10 percent, the total number of required samples $n = (8\sigma/\Delta)^2$. Here σ is the standard deviation

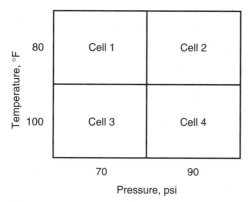

FIGURE 7.25.   Cells for the experiment.

of the process and Δ is the shift required to declare the new process better than the current one.

Suppose that a shift of 2 σ would qualify as an improvement. The sample size, after substituting the values in the preceding formula, would then be 16. Since there are 4 cells, the required number of samples per cell would be $16/4 = 4$.

Perform the experiment per the conditions specified in the respective cells, and complete the whole process cycle four times under each condition. In performing the experiment, it is important to perform the cycles in random order, so that the outcomes are not affected by the cyclical nature of the process or some other trend that might cause an erroneous output. For this purpose, the sequence may be selected with the help of a random number table.

Record the Outcomes.   Assume that the outcomes of the experiment are as shown in Fig. 7.26.

Step 5.   *Analyze the Data Interactions.* An interaction occurs when the effect of one input variable on the process output depends on the level of another input variable. Interaction is one of the most important gifts of factorial design. It displays the simultaneous effects of the variables on the process. Interactions are the driving force in many processes. Figure 7.27 elaborates this concept.

Note from the first diagram in Fig. 7.27 that the value of output at high or low pressure moves in the same manner irrespec-

	70	90
80	119 110 115 120	90 75 100 108
100	100 102 116 115	140 125 130 110

Temperature, °F

Pressure, psi

FIGURE 7.26.   Experiment output.

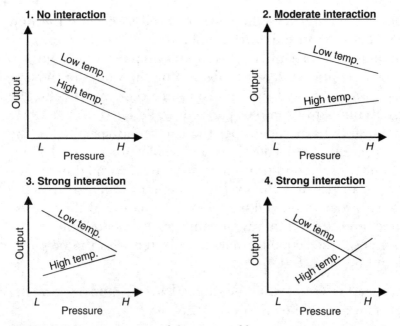

FIGURE 7.27.   Interaction of the two variables.

tive of the temperature level—that is, the gap between the outputs does not vary with the temperature. However, in the second diagram, the gap between high-temperature and low-temperature output varies with the level of pressure, and this shows that something is going on in the process that is changing the gap. In the third and fourth diagrams, this gap varies sharply with the level of pressure; hence, the interaction is very strong.

In order to observe the interactions, you need to plot the means. The mean of the observations for each cell is calculated, and the results are displayed in Fig. 7.28. The interactions are displayed in Fig. 7.29.

Note that as the pressure increases, the output of the process at low temperature decreases, while that at high temperature increases. This inverse movement of the outputs confirms the strong interaction of the processes. We will examine this phenomenon in detail shortly.

Step 6.   *Analyze the Data—Table.*

After you have seen the type of interaction, you need to look at the effects of various variables and their interaction on

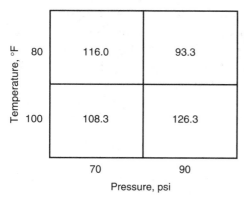

FIGURE 7.28.   Mean of the output data.

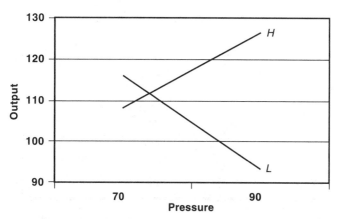

FIGURE 7.29.   Plot of the mean of the output data. Lines plot temperature.

the output, as presented in Fig. 7.30. Follow these steps to arrive at the values in the figure:

1. Pick the data from Fig. 7.26.

2. For each row, representing the cells, add all the observed values. For example, the total for Cell 1 would be 119 + 110 + 115 + 120 = 464, and so on.

3. For each column, representing variables and interaction:
   Pick and add all H values for each variable and interaction from the total for the corresponding cell calculated in Step 2, irrespective of the level of other variables and interactions. For example, for the temperature column, the total would be 433 (Cell 3) + 505 (Cell 4) = 938.
   Next, pick and add all L-values for each variable and interaction from the total for the corresponding cell calculated in Step 2, irrespective of the level of other variables. For example, for temperature, the total would be 464 (Cell 1) + 373 (Cell 2) = 837.

4. Now calculate the effect of each variable and interaction, as follows:

$$\text{Temperature effect} = \frac{938 - 837}{8}$$

CELL	TEMPERATURE	PRESSURE	INTERACTION	TOTAL OUTPUT
1	L	L	H	464
2	L	H	L	373
3	H	L	L	433
4	H	H	H	505
Sum of all H	938	878	969	
Sum of all L	837	897	806	
Difference	101	−19	163	
Effect	12.6	−2.4	20.4	

FIGURE 7.30. Data analysis matrix.

$$= -12.6$$

$$\text{Pressure effect} = \frac{878 - 897}{8}$$

$$= -2.4$$

$$\text{Interaction effect} = \frac{969 - 806}{8}$$

$$= 20.4$$

Note from the matrix that the effect of pressure is in the reverse direction (-4.8); that is, the output decreases with increased pressure. The effect of temperature is very strong (12.6) on the output; that is, the output increases with increased temperature. The effect of interaction on the process output is significant—a sharp increase (40.8) in the output as the level of interaction becomes high. The interaction effect on this process is much higher than the effect of any input variable.

Now you can say from the preceding analysis that the best way to run the process is to keep both the temperature and the pressure high; that is, operate at Cell 4. This would keep the process at high interaction $(+20.4)$, at high pressure (-2.4) and at high pressure (12.6). Note that although the effect of temperature is negative, if you keep the pressure low and the temperature high, the interaction is low and the overall effect is worse. Hence, the experiment indicates that the following parameters improve the process:

$$\text{Temperature} = 100°F$$

$$\text{Pressure} = 90 \text{ psi}$$

In order to confirm that this is a statistically significant change with respect to the current process, it is a good idea to conduct two sets of experiments to perform current-versus-better analysis.

ROBUSTNESS—OPTIMIZATION

Robustness is the concept of producing an output that is insensitive to the sources of variation of the inputs. The need for this concept arises from the fact that customers feel the variance and not the mean output. Consistency of output is achieved only when the variation is reduced. This can be achieved by moving the mean performance to the target as well as by decreasing variation around the target.

TAGUCHI METHOD

The Taguchi method, developed by Genichi Taguchi, refers to techniques of quality engineering that embody both statistical process control (SPC) and new quality-related management techniques. The prime motivation behind the Taguchi experiment design technique is to achieve reduced variation. This technique, therefore, is focused on attaining the desired quality objectives in all steps. The classical DOE does not specifically address quality; it rather assumes it to be given while performing the experiments.

The fundamental philosophy of the Taguchi method can be summarized in the following two basic ideas:

1. Quality should be measured by the deviation from a specified target value, rather than by conformance to preset tolerance limits as is done in the traditional statistical process control charts.
2. Quality cannot be ensured through inspection and rework, but must be built in through the appropriate design of the process and product.

The first concept underlines the basic difference between the Taguchi method and the SPC methodology. SPC methods emphasize the attainment of an attribute within a tolerance range and are used to check product or process quality. According to SPC, the product just inside the tolerance limit is good and the one just outside the limit is bad. Conversely, the

Taguchi method emphasizes the attainment of the specified target value and the elimination of variation. According to this method, there is no significant difference between products that are just outside or just inside the tolerance limits. This concept is illustrated in Fig. 7.31.

In the second concept, the Taguchi method emphasizes that control factors must be optimized to make them insensitive to manufacturing and input variations through design, rather than by trial and error. SPC allows faults and defects to be eliminated after manufacture, whereas the Taguchi method prevents their occurrence. The objective is to achieve consistent performance by making the product or process insensitive to the influence of uncontrollable factors. Through the proper design of a system, the process can be made insensitive to variations, thus avoiding the costly eventualities of rejection and rework.

Quality Loss Function.   Various costs associated with any product, service, or process are as follows:

Design costs.   Design costs are incurred in the development of new products and the study of existing products with respect to the ever-changing customer requirements.

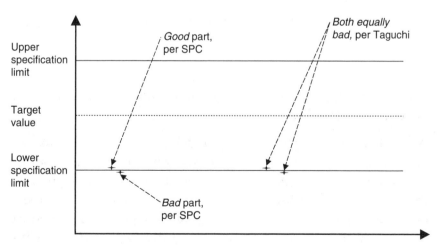

FIGURE 7.31.   Comparison of the traditional SPC and Taguchi methods.

Manufacturing costs. Manufacturing costs include total inventory costs, machinery costs, labor costs, and costs of all the waste generated in the production unit.

Sales and distribution costs. Sales and distribution costs are incurred after manufacturing the product. They include finished goods inventory costs, promotion costs, obsolescence costs, and warranty costs.

Operating costs. Operating costs are incurred in operating the product by the organization's customers. They include energy costs, additional training costs, environmental costs, and other special considerations (such as air conditioning) required to operate the product.

In this competitive environment, it is expected that the organization of the future will not only be sensitive to all these costs but also will make a conscious effort to minimize each one.

According to Taguchi, the deviation of quality from the target value results in a loss to the society as a whole. This loss consists of opportunity costs and wasted resources and time. This loss is proportional to the square of the deviation from the target and may be expressed as follows:

$$\text{Loss} = k(y - m)^2$$

where k = quality loss coefficient (constant)
y = actual value of output
m = desired target value

The implications of this equation with regard to loss are depicted in Fig. 7.32.

Noise Factors. It is important to select a product design or a manufacturing process that is insensitive to uncontrolled sources of variation. It obviously will improve the quality. Taguchi calls these uncontrolled sources of variation *noise factors*. The purpose of identifying the noise factors is to make the products insensitive to their influence. Noise factors are those factors that are not controllable, whose influences are not known, and which are

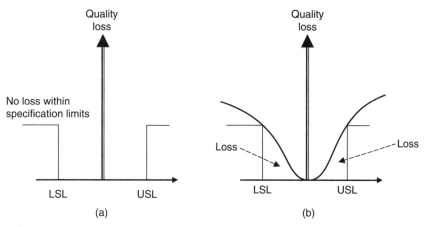

FIGURE 7.32. Comparison of loss: (*a*) traditional SPC method, and (*b*) Quadratic loss function.

intentionally not controlled. To determine robust design, experiments are conducted under the influence of various noise factors. The noise factors are broadly classified as follows:

External noise factors.   External noise factors are the conditions under which the product operates. These factors include:
- Environmental temperature
- Humidity
- General hygiene
- Operator error
- Duration of operation

Product-to-product variation.   The product-to-product variation is primarily due to the accumulated effect of the tolerances specified in the inputs and the process conditions.

Degradation over time.   Degradation of the product occurs over a period of time due to the limited life of the product's constituents. This is quite possible even when the product meets all specifications on leaving the factory.

Producing a Robust Design.   The Taguchi method is used to improve the quality of products and processes. Improved quality results when a higher level of performance is consistently obtained. The highest possible performance is obtained by determining the optimum combination of design factors. In the Taguchi approach, optimum design is determined by using DOE principles, and consistency of performance is achieved by carrying out the trial conditions under the influence of the noise factors.

In order to determine and subsequently minimize the effects of factors that cause variation, the design cycle is divided into three phases: *concept design, parameter design,* and *tolerance design.*

Concept Design.   Concept design focuses on the very beginning of the conceptualization of the idea to make the product and the process. The key consideration during this phase is how to minimize the manufacturing costs.

Parameter Design.   Parameter design focuses on the level of process parameters for optimal process performance. The ideal system is one that minimizes quality loss with more economic input substitutes. Such a design is illustrated in Fig. 7.33.

As you can see from Fig. 7.33, Output 1 has a very wide range for a relatively low variation of Input 1. But when operating at Input 2 levels, the range is quite wide but the Output 2 range is very narrow—that is, the sensitivity to the noise has decreased in this case. This implies that as far as possible you need to adjust the input parameters in such a fashion that the output is relatively insensitive to the variation of the inputs.

Tolerance Design.   After concept design and parameter design, you know you have the right product and parameter designs. Now you need to implement them in the place where the real action is—the shop floor. The tolerance design is a compromise between the high manufacturing cost associated with lower tolerances and the poor quality associated with wide tolerances.

RESPONSE SURFACE METHODOLOGY

After conducting DOE and process modeling, the knowledge gained identifies the root causes of a process problem or criti-

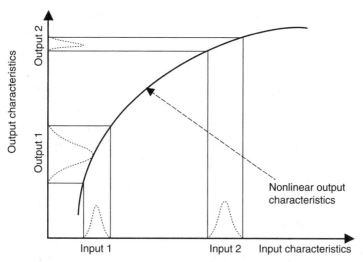

FIGURE 7.33.   Exploitation of the nonlinearity of the output.

cal variables related to the process yield. One can be satisfied with a DOE leading to some improvement, but to achieve the best yields and reduce the total cost, one must look into optimizing the process. The total cost includes the following:

• Doing the work (setup and operation)
• Purchasing material and supplies
• Purchasing equipment and tools
• Moving inputs and outputs
• Inspecting, testing, and verifying
• Sorting, repairing, or reworking
• Expediting
• Transporting

Optimizing a process is an economic decision, overlooked by engineering and management in many companies. As processes become complex, thereby increasing the process cost, one must look into optimization instead of maximization. Optimization implies working with multiple, sometimes conflicting, variables at the same time to improve a process.

Ideally, one would like to optimize a product or process during the design phase when it is being qualified for reproducibility. Then, if the variables are extended to the worst-case process parameters expected, one could anticipate and reduce potential problems in the production or postdesign phase. Optimization can be thought of as building a process model and finding a zone that is insensitive to input variation. Optimization is a "best" decision at a given time. One cannot just go wrong. While seeking breakthrough solutions, one must not ignore by-products of the dramatic solution.

Response surface methodology (RSM) is an experimental approach to fine-tune and optimize the process, or achieve the state of robustness. RSM is a tool to find a zone where the variability in the process output is minimal. RSM is a combination of DOE and process modeling using regression analysis techniques. One can look at RSM as a process characterization method. The benefits of RSM are as follows:

- Improved yields and lower defect levels
- Less process variation and more predictable performance
- More predictable process cycle times and reduced lead time
- Reduced inspection, testing, and verification requirements
- Lower product or service cost

When optimizing a process, RSM works by exploring the operating range of the process in the direction with the highest rate of improvement, or the steepest slope. The method is called the *path of steepest ascent* (PSA)—that is, the parameters are varied in the direction of maximum increase of the response until the response no longer increases.

For example, say that process yield depends on the temperature. Current understanding suggests that as the temperature rises, the yield increases. However, if you explore the relationship further, you will find that there will be a temperature beyond which the yield starts to decrease. In other words, the point of diminishing returns is reached. This is illustrated in

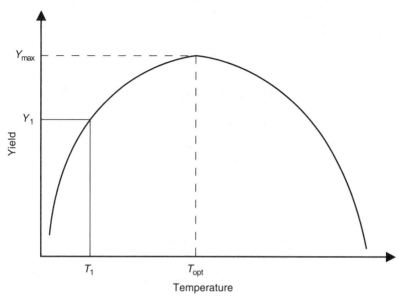

FIGURE 7.34.   Response surface methodology.

Fig. 7.34. RSM is used to identify the region where the yield is highest and to determine the range of input variables to maintain the yield at the highest level consistently. This region is investigated in more detail with smaller increments using software.

RSM Guidelines. RSM utilizes many DOE techniques. Besides, process knowledge and engineering judgment are critical in optimizing a process rapidly. The mechanical aspects of RSM such as DOE, calculating the PSA, and fitting a model cannot be accomplished without the help of statistical analysis software. However, a successful RSM must follow a process approach to investigate the process optimum. Questions such as which variables to look into, the operating range of various variables, sample size, expected process outcome, incremental steps of experiments, and the direction of continual experimentation are addressed in RSM.

RSM is a data-driven approach to process output optimization. It starts with the knowledge of the few most important

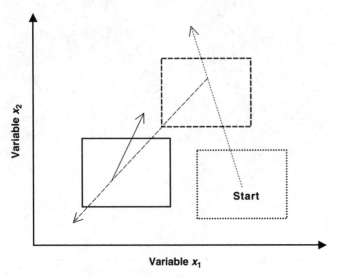

FIGURE 7.35.   Following PSA to locate Optimal Output.

process variables, which have been determined using prior planned experiments, such as full-factorial experiments. RSM helps you move rapidly toward the neighborhood of the optimum output by performing a series of simple experiments, which identify the PSA.

SUSTAINING BREAKTHROUGH— CONTROL PHASE

Having found a solution for the problem and implemented it, the challenge is to keep the gains and sustain improvement. In today's competitive environment, continual implementation of Six Sigma or reengineering of processes is mandated. In order to control gains and sustain improvement, one must ensure that the right thinking (statistical, process, and innovative thinking) about the process is inspired. Implementing the control phase is beyond statistical process control (SPC). Instead, one must make sure that the intent of processes and business objectives is understood that leadership continually inspires improvement, and that measurements are maintained to assess improvement and give feedback. The following elements constitute the activities required to implement the Control phase or sustain the breakthrough improvement:

1. Thinking right
2. Managing the process
3. Valuing workers
4. Leading improvement
5. Counting the change

THINKING RIGHT

All business endeavors consist of a collection of processes with identifiable suppliers, customers, and sets of requirements on all the deliverables involved. The challenge of managing the Six Sigma initiative comes from the variance that occurs in different processes, in the goals of the functional owners, and in the overall business goals and objectives. The overarching directional control normally comes from the marketplace, the competition, and business leadership with the good business sense to align with this direction. As the team has implemented the Six Sigma DMAIC methodology to improve the business process, it has achieved some notable success in the earlier steps toward achieving the goals of the organization. These steps might include identifying customer needs, available methods and technologies to create these customer-satisfying values, and processes to be followed. The team has also created a new process model, tested it, validated the process, and implemented it. The steps perhaps include formal and informal training, testing, evaluation, and even certification.

When these improvements have been implemented, there may be a tendency to say "well done" and move on to the next thing. This impulse to keep moving on to the next project without seeing successful implementation and control in operations must be restrained. The DMAIC process needs to be properly completed, with the implementation of *control* steps that ensure the continual success of the improvement effort and provide feedback and monitoring mechanisms that make the process alive and organic.

DEFINING ROLES AND RESPONSIBILITIES

One of the key steps in control involves documenting the responsibilities and accountabilities of every employee, including the management and leadership involved in reaching the Six Sigma project objectives. The main purpose of documentation is to maintain consistency and control in the process operation. Consistency relates to compliance, and control relates to effectiveness. This also means that doing the right thing involves doing things right, creating the opportunity for the business to be both effective (doing the right thing) and efficient (doing things right). This moves the business from a *product* focus to a *process* focus, and this essentially removes the difference between employees and management and shines the spotlight on the process. Everyone, including those in management, must perform their part of the tasks to meet the intent of the quality processes. This creates synergy and harmony, and the quality management system facilitates growth, downsizing, and regrowth equally well.

DOCUMENTING PRACTICES AND CHANGES

Controlling the improved process requires that the knowledge gained by applying the DMAIC methodology and revising the process for improved performance must be captured to prevent its loss and to propagate it farther. Documentation brings the clarity of a program of steps to be taken to sustain the gain realized.

When documenting a quality process, the goal must be to document the process meaningfully, to serve its intent. The process details must be based on the competency of the users rather than on the software used or the author doing the documentation. Otherwise, the company accumulates documentation of little value—too much of a burden to read, understand, and maintain. During the documentation process, obtain everybody's buy-in by giving ownership to the employees. The key to the success of this approach is in setting the metrics for deliverables to ensure meeting or exceeding customer expectations. As an example, a voluminous product document consisting mostly of descriptive how-to procedures with very few illustrations can be replaced with a few pages of progressive illustrations based on the needs of the users.

Another important function served by documentation is the ability to find the right root cause of problems with ease. Without documentation, it is difficult to identify consistent practices and the causes of variation in them. Without documentation, it is impossible to identify the causes of problems in a process with confidence. This leads to treating the root causes effectively instead of treating only the symptoms, which could be a total waste.

PROCESS MODEL FOR DOCUMENTATION

Generally, a good procedure must address the elements of a process model, as shown in Fig. 8.1. Essentially, the process model consists of Plan, Do, Check, and Act (PDCA) activities. According to the process model, any activity, or set of activities, using resources (capacity, staff time, supplies, energy, facilities, etc.) to transform inputs to value-added outputs could be considered a process. The efficiency of this transformation of input to output depends on identifying and managing various processes and their interactions under a quality management system. In order for the quality management system to be applicable companywide, deployed process management at each process becomes a precondition. The process owner needs to understand the requirements, and needs to manage the receiving, processing, and supplying of the deliverables according to those requirements. Compliance at all stages is achieved by verification methods. If verification shows that requirements are not being met, corrective actions are launched to bring the process under control. Once the process is managed effectively, one can identify critical elements in the organization or in the process for control activities.

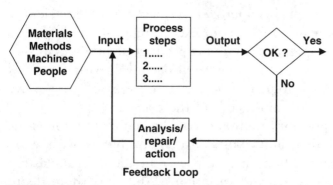

FIGURE 8.1.   Process model.

Training

Training has been found to be the best investment for a corporation. When a new methodology or a process change is introduced in an area, training is the most critical aspect in the successful implementation of that change. Training can be used to communicate changes, to overcome initial resistance, to gather feedback, or to transfer the knowledge necessary to control the processes. Training can be performed in a classroom environment or through electronic media. When planning for training, one must ensure that participants learn by practicing in the classroom and understand the intent of the training. In sustaining the benefits of a Six Sigma project or initiative, participants must understand their ongoing roles and the activities to be performed consistently. Any plan for training must also include management personnel at the appropriate level.

A new process, when implemented, must affect the organizational culture positively. Employees must be able to see its benefits as well as the challenges in implementing the process. Sometimes multiple departments—sales, design, purchasing, manufacturing, services, and management—are affected. The benefits must be significantly greater than the pain caused by the process change. The benefits may include the following:

- Better understanding of the commitment to improve
- Better knowledge of the process and its intent
- Clear understanding of expectations
- Better knowledge of how to do a task consistently
- Better knowlege of how to adjust a process if the desired performance is not realized
- Clear understanding of who to get to help when needed to control a process

Training prepares employees to implement the process change correctly and to maintain the process at the high-performance level. Training can be performed at multiple levels—detailed operational training for operator level, technical and maintenance training at the engineering level, or an overview at the management level. The following lists the key aspects of successful training:

- Focuses on understanding the intent and requirements of the process
- Explains the roles and responsibilities of management and executives
- Recognizes that awareness training is more fundamental and powerful than tools training
- Assesses the effectiveness of the training based on participants' feedback

MANAGING THE PROCESS

Thinking right requires a good understanding of the process model. Managing the process requires good process controls, ensuring that appropriate remedial actions based on facts are taken. For good process controls, one needs good and reliable data, timely interpretation, and appropriate response. In the absence of useful data, feelings and emotions run the business; management is burdened with decision making, employees are blamed for mistakes, quality is measured by the number of quality inspections, and performance is measured by customer complaints. The quality department is blamed for not catching all the defects. Customers are gained based on the sales price. The stakeholders' mission becomes getting out of the business profitably, instead of growing the business profitably. The chain of activities (including customer complaints, returned material, and corrective action requests) and the root-cause analysis still take place; however, finding a solution becomes difficult. In order to continually apply DMAIC and Six Sigma principles, one must manage processes well, through good data collection and effective implementation of control charts.

Data Quality

Invariably, companies collect data about every process they possess. Even when the data are collected religiously, nothing happens with the data, and no return is realized from the effort. Sometimes, when a company has implemented some analysis procedures, the data are fake, incomplete, inaccurate, and, most important, unreliable. Sometimes the data are not even appropriate, and still no one asks questions. In such cases, if someone throws data out the window, no one bothers to look for them. Actually, that is what happens in many companies where no one knows where the data disappear to after they are collected. The main issue with the data is that tons of data are collected, but none of them are analyzed. Even when the analysis takes place, it is shallow and superficial, as the analysis does not identify new opportunities for improvement. People look at the data for a few seconds instead of looking into the data for a while to extract information.

Many in industry have not understood the link between data collection and value to the business. The ISO 9000 requirements prompted many companies to create documented data collection procedures. Both the consistency of data collection and timely, critical analysis for the information and knowledge content are critical to the profitability and success of the company. To be consistently profitable, the business must know how well its operations are running, what the yields are, what the major problem areas are, and, most important, which process variations need to be reduced. Excessive variation will lead to rejects, a cost that adversely affects the profitability and the long-term success of the business.

Six Sigma is a fact-based or data-driven methodology. Decisions are made based on the data. Normally, the data are collected during the project life, instead of under normal operational conditions. If the data are not collected under normal working conditions, the Six Sigma project team has a tough time finding data. The team must collect the data in order to make statistically correct decisions. At the time, this appears to be an extra task. The new quality standards emphasize data analysis in critical areas for process and business improvement. Without proper emphasis on the collection of data to measure the effectiveness of processes, it is difficult if not impossible to create value. This is because managers make decisions primarily from personal memory or convenience in the absence of good data collection and analysis. Without data analysis, there are no visible opportunities for improvement that could lead to process improvement. In other words, no data collection means no problem identification; therefore, no process improvement; and finally, no improvement in profitability. It is a struggle everywhere, inside and outside the company.

To sustain the process improvement realized through the Six Sigma project, one must establish the process model for the process and identify critical process parameters at input, in process, and at output. Then, establish appropriate measurements, such as DPU, DPMO, C_p, or C_{pk}. For attribute data, one may establish DPU, while for variable data one may establish C_p and C_{pk} measurements. After identifying the measurements, and based on their capability, one can decide what type of control charts or equivalent tools to use.

The event cycle runs from collecting data to controlling the processes within operating specifications, learning about process performance, and identifying opportunities for improvement. This sequence of activities also provides information for the use of leadership in long-term planning and capital investment. The main purpose of data collection and analysis is to make the process problems visible so they can be addressed, not ignored. Usually, a problem not seen is a problem ignored. A problem that is known has potential for being addressed, and a problem well defined is half solved. Problem definition can happen only with data collection and valid analysis for meaningful insights into the process. It is almost an extension of logic that if there is no data collection in a business, it is likely that the business's reject rates are higher and that the root causes of problems are seldom identified or addressed. For an operation without data, the quality level can be expected to be around two or three sigma—and the business can be labeled an endangered corporation.

Data collection must address the following points:

- Was the process under control?
- What is the capability of the process with respect to the specifications?
- What is the typical process performance?
- What is the process problem?
- What is the magnitude of the problem?
- How frequently does the problem occur?
- What were the process conditions at the time when questionable parts were produced?

Data Collection Plan. A valuable data collection system is a planned effort by management and employees. A good data collection system is defined with a purpose, scope, and usage. The monitoring of a process must necessarily look at all the inputs to the process, in-process parameters, and of course the output. The monitoring of a product may concentrate on product inspection or test data.

The purpose of the data collection must be clearly specified and understood. Once the purpose is established, methods, tools, and plans to collect the data are developed. To maintain the benefits of the solution developed by the Six Sigma project team, new process parameters must be specified and used for data collection. Accordingly, a new log sheet or check sheet is used. A good data collection log sheet is designed with consideration of the process flow or sequence of operations, such that the information is entered on the sheet in the order in which it is collected. The check sheet could also graphically summarize the data as it is being collected. Irrespective of the design, there must be clear instructions for the operator regarding what data to collect, when to collect, how to collect, what information to enter on the form, and what resolution of the data should be recorded on the form. This is an issue that has been noted in the industry. Without a proper rationale or instructions, the measurements are rounded off, thereby reducing the resolution of the measurement system, rendering it incapable of capturing process variation. Without proper resolution, the data will show the average but not the variation, whereas most problems occur due to variation in the process, not in its mean.

Training for Data Collection.   Once the parameters are identified and the tools are designed, people must be trained to collect accurate data. Before implementing the new data collection method, it is imperative that someone communicates with the people who will be collecting data. No matter how commonsense it may appear, people forget to communicate instructions or train people in new data collection methods. As a result, when the new method is expected to be working, people in operations may not be even aware of its existence. There must be a clear switchover or starting date for the new data collection method. After some time, one may conduct audits of the data collection process during the regular process audits, or just for data collection companywide.

If management is not involved in the emphasis on good data collection, expectations are set for not paying attention to either the process or the data. The only way to align the process, data collection, and management's attention to them is to use the facts based on the collected data in making decisions on which the profitability of the business depends. Data are the voice of the process that management needs to hear, understand, and act on as needed. Failure to do so misses the opportunities for improvement, a vital task in managing a process. The data on shipment are important; so also are the data on total production, waste, rejects, and every other category, because customers pay not for producing, but for producing *good product*. The collected data about the quality of the process and the product, and about customer satisfaction, are the mainstay of process analysis, control, and continuous improvement to sustain the company and improve profitability going forward.

CONTROL CHARTS

Control charts are a widely used tool for maintaining processes in statistical control. Control charts have been misunderstood and misused for a long time. People have produced and posted control charts to demonstrate their commitment to SPC when their customers have demanded it. Typically, when push comes to shove, someone pushes for control charts, they suddenly appear all over—and then disappear just as rapidly, too. One of the main requirements for implementing control charts is that the process must be in statistical control. However, in most cases, control charts are implemented to try to bring process under statistical control. The problem with this approach is that the process shuts down too frequently, which is intolerable to plant management. So, the first thing that happens is that the control limits are removed. The ritual of plotting data then continues without any limits, so there are no out-of-control situations, no adjustments to the process, no reactions based on the control chart, no process activity—and, therefore, no interest. Since the charts add no value, employees become disinterested in them. They stop plotting the data, management does not expect the data, and the employees find out about it. So they stop collecting the data altogether, until the customer next pushes the panic button.

All of the abuse of control charts occurs because there is a lack of understanding of their intent and purpose of the nature of variation. Every process has its central tendencies and tail ends, due to inherent variation in the process. When a process varies from the target or its natural pattern of variation, it incorporates nonrandom variation, which we have understood to be the *assignable* variation resulting from some assignable cause. Any variation from the norm costs money in inspection, testing, and verification. When the variation becomes excessive, beyond the specification limits, rejects are produced. All quality systems address this aspect of variability, to minimize or eliminate this cause when possible. On the other hand, when random variation in a process is unacceptable, it must be reduced through process capability studies and improvement activities.

Walter Shewhart, working at Western Electric's Hawthorne plant near Chicago, Illinois, developed control charts in 1929 and initially published his findings in 1931, under the title

Economic Control of Quality of Manufactured Product.
Shewhart identified variation as the enemy of quality. This core
message subsequently formed the basis for W. Edwards
Deming's management philosophy on quality. Deming called
variation "evil." In order to implement statistical control and
use statistical techniques, understanding statistical thinking,
or the difference between random and assignable variation, is
a must.

Definition.   The control chart is a tool used to evaluate variation in a process and determine whether the variation is random or assignable. In other words, is the variation common,
usual, normal, or typical? Or is it assignable, special, rare,
nonnormal, or atypical? The control chart compares variation
in a process with properly established control limits based on
the inherent variation (historical variation from a larger data
set) in the process. If the variation follows the probability law
of normal distribution, the variation is random and therefore
must be left alone. The control chart indicates when a process
has exceeded the control limits and has been introduced by
some assignable cause. If the process appears to be out of control, the process must be adjusted. The adjustment may involve
doing nothing; shutting the process down; or quarantining,
sorting, or inspecting the production. Shewhart said that control charts (or any other tool) must not be implemented unless
it makes economic sense to do so.

Simply put, a control chart is a device that determines
whether the process is within the bounds of statistical control.
If it is not under statistical control, then action must be taken
to bring it back under statistical control before resuming the
process.

Normal Distribution and Control Charts.   When a process
is under statistical control, its output follows a normal distribution. When a process experiences an assignable cause, it loses
its normality. Therefore, one can say that the control chart tests
for the normal behavior of the process. To enable this, the control chart rules are based on the normal probabilities. Control
charts tell you whether the probability of producing a product

using a given process follows the normal distribution. Therefore, a control chart evaluates variation in a process for normal distribution around a target (typical value) and control limits (acceptable variation from the target). When the variation exceeds the probability of producing a product at a target value (0.27 percent or less), the cause of the variation appears to be identifiable, or is considered to be *assignable*. When the variation is such that the probability of producing a product at a target value is more than 0.27 percent, the cause of the variation becomes difficult to determine and is considered *random*.

Preparation for the Control Chart.   The control chart must be used when a process has been properly characterized in terms of its statistical variation. All the assignable causes must be removed before using a control chart to monitor a process. In reality, the charts are often established with assignable causes present in the process. If assignable causes are present, the process will have to be shut down frequently, and the chart will become a nuisance to management and be ignored by the operators. Therefore, control charts must be established when the process is under statistical control, that is, when only random causes are present. No known problem should exist that could cause variation beyond the control limits.

Types of Control Charts.   Control charts are classified into two categories, based on the type of data used for controlling the process. *Attribute* control charts are used when attribute data is collected, and *variable* control charts are used when variable data is collected. Examples of variable data include time, height, thickness, temperature, pressure, current, voltage, and power. Examples of attribute data include good or bad, pass or fail, high or low, and accept or reject. Variable-data charts are used for processes that are complex and continuous in nature, while attribute-data charts are used to control processes that are mechanical and less complex, such as assembly processes. Various types of control charts are as follows:

Attribute Control Charts

u Used for plotting the average number of defects per unit in a subgroup *np*   Used for plotting the number of defective units in a sample for go and no-go data

p   Used for plotting the percentage or proportion of defective units in a subgroup

c   Used for plotting the number of defects in a subgroup

Variable Control Charts

$\overline{X}/R$ Used for plotting the mean and range of subgroups of greater than 1 and usually less than 5

$\overline{X}/s$ Used for plotting the mean and standard for a larger sample size 5 or greater

$\overline{X}/R$ (or *MR*)   Used when measurements are taken on individual units

Regardless of the type of control chart, the basic rules for maintaining statistical control are the same. The rules are based on the probability of having data points in the area between the number of standard deviations and the mean. The main difference is the method used to calculate control limits. The methods use different formulas. Initially, one can try to calculate manually and make a chart. In the long term, the limits can be calculated using statistical quality control software.

Constructing a control chart requires decisions about control limits (upper and lower), sampling frequency and sample size, central tendency, and acceptable variation. Besides, action plans to address out-of-control conditions must be understood and agreed upon. When developing a control chart, one must consider the inherent limits of the process, which dictate the parameters. This also implies that a process to be controlled needs be free of known limitations and defects that can be fixed by routine repairs, maintenance, or upgrades. Otherwise, you may be creating a control scheme with built-in limitations (known causes), leading to poor control.

Constructing the Control Chart

- Select process parameter or characteristics.
- Develop the check sheet to collect data.
- Collect samples and record the data.
- Calculate the subgroup averages.

- Calculate the process mean.
- Determine the ranges for the samples, and calculate the average range.
- Determine control limits for the averages and ranges.

Alternatively, one can collect the data and feed them to a computer, which will spit out the control charts promptly.

Before implementing a control chart, one must examine the process for root causes of potential problems that could lead to out-of-control situations. This is another way of removing any assignable cause that may exist in the process. One really must start using a control chart to control a process after the process has been cleared of any assignable or known problem or cause. One can ask the following questions to investigate for existing assignable causes:

- Is the right equipment being used?
- Is the equipment functioning properly?
- Is the equipment being operated correctly?
- Is the material within specification?
- Is the process set up correctly?
- Is the process effective to its "should-be" level?
- Are employees qualified for the process?
- Were the employees trained to correctly use the control charts?
- Is an escalation process set up to ensure correct implementation of the control charts?
- Are the support personnel trained in maintenance or calibration?
- Is there documentation of solutions to past problems?

Interpreting the Control Chart.   Figures 8.2 and 8.3 show an example of a control chart and equations for setting control limits. Correct interpretation of control charts is necessary to adjust the process appropriately. This is the weakest

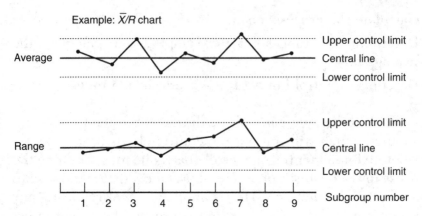

FIGURE 8.2. Example of a process control chart.

Control Limits	Factors for Control Limits		
	n	A_2	D_4
$\bar{\bar{X}} = \Sigma(\bar{x})/k$	2	1.88	3.268
$UCL_x = \bar{\bar{X}} + A_2\bar{R}$	3	1.023	2.574
$LCL_x = \bar{\bar{X}} - A_2\bar{R}$	4	0.729	2.282
$UCL_r = D_4\bar{R}$	5	0.577	2.114
	6	0.483	2.004

Where k = number of subgroups,
and n = the subgroup size

FIGURE 8.3. Setting control limits.

link in using the control charts. Standard rules for interpreting control charts are widely known and should be used to determine whether the process is under statistical control or out of control:

- Data point beyond control limits
- 9 points in a row on one side of centerline
- 6 points in a row increasing or decreasing
- 14 points in a row alternating up and down
- 2 of 3 points in a row beyond 2 σ
- 4 of 5 points in a row beyond 1 σ

- 15 points in a row within 1 σ
- 8 points in a row on both sides of the centerline within 2 σ

In following these rules, one can see that the intent of the control chart is to test for probabilities according to the normal distribution. If points fall beyond the control limits, there are several possible causes. The control limits may be in error, the process performance may have changed, or the measurement system may have significant variation. If patterns or trends exist within control limits, this may indicate an out-of-control condition; that is, some known or assignable cause may have been introduced in the process. Runs (points lined on one side of the process mean) could indicate a shift in the process mean. Any nonrandom pattern indicates the presence of an assignable cause, requiring immediate attention. One can formulate responses based on the analysis of charts as follows:

- *Points beyond the control limits.* Limits may be in error; process performance has changed, or the measurement system has changed.
- *Patterns or trends within control limits.* These can also indicate an out-of-control condition.
- *Runs.* These can indicate a shift in the process mean.
- *Obvious nonrandom patterns.* These indicate the presence of assignable causes, requiring immediate attention.

Overreacting to the data is just as detrimental to process control as not reacting at all. Control charts prevent overreaction as well. Sometimes people react to any variation from the target and adjust the process right away. One must recognize normal variation and leave the process alone. Even in the presence of an assignable variation, adjusting the process back to the point of inflection may be an overadjustment. Instead, a good adjustment would be to adjust the process mean either back to the target value or halfway toward the last good data point.

VALUING WORKERS

With today's global and diverse workforce, competitive environment, and increasing customer demands, the work environment is changing. Work no longer means a physical job; instead, it is becoming more critical that employees are motivated to be intellectually involved. The role of leadership is evolving, such that instead of the workers doing perfect work, it is the leadership that must do perfect work. Among the first and foremost responsibilities of leadership are to inspire workers to do their best, to value their contributions, and to create new opportunities for innovative solutions.

To develop an innovative product, service, or solution, workers must be fully engaged and given the responsibility for meeting business objectives. Controlling a process is not enough; instead, controlling the implementation of breakthrough or innovative solutions has become a critical aspect of the Six Sigma initiative. Whether in the service industries, manufacturing, or the software industry, in both large and small businesses, innovation is mandatory. Innovation can occur when employees feel their ideas and recommendations are valued. They bring out more ideas; one good idea breeds more good ideas. Employees need to know that their ideas are good ideas.

Valuing workers requires implementing leadership process to bring out their best, and giving employees the responsibility and authority to manage their processes. Precontrol charts are a tool for controlling more mature processes and giving process operators the authority to manage them.

LEADERSHIP FOCUS

The company leadership must focus on adding value for the customer rather than shipping the least expensive product. When the focus is on price, the supplier has no other opportunity to add value; instead, the supplier focuses on shipping a product that is feasible at the lowest cost while still meeting the customer requirements. Customers have been asking for better, faster, and cheaper products or services; however, experience proves that they really want cheaper, better, and faster, or cheaper, faster, and better. In the service environment, it is faster, cheaper, and better, because the definition of quality is not so clear. Customers must understand the *total* cost of purchasing instead of just the *initial* cost. The total cost includes product cost, freight charges, customer service cost, cost of communicating requirements or changes in requirements, on-time delivery cost, cost of handling rejects, and cost of dependability. Leadership can play a role in establishing value- and quality-based relationships with customers that require constant communication with customers and employees.

So what can the supplier's leadership do to change the focus from price to quality and performance? The leadership must focus on and commit resources to improving quality. Any improvement in quality will improve the cost, on-time delivery, shipment, waste, and profitability factors. Due to the continual pressure on the profit margin, improving processes becomes the number-one priority. One can learn from the automobile industry, where the customer, through a QS-9000 quality system, requires suppliers to set continuous improvement goals for quality, productivity, efficiency, downtime, and so forth. There the intent is to reduce the cost of materials and the cost of manufacturing to stay competitive in light of the fierce competition from offshore suppliers.

One of the major concerns in many industries is the skill level of the workforce. In addition, due to advances in technology, the requirements for employee skills have never been so high. In the early 1980s, companies such as Motorola, Xerox, Texas Instruments, IBM, and Hewlett-Packard implemented some basic skills programs to improve employees' reading and math skills. Government supported the effort by sharing the cost of training through state grants. One should look for some government help in expanding employee skills. Business processes are being outsourced because, based purely on wages, it is becoming difficult to compete globally. However, the skilled workforce can innovate better solutions, adding value and service for customers, who would then understand the total cost of doing business, not just the initial cost. Customers, suppliers, and the community must support skill development programs. Only a trained and skilled workforce will be able to keep up with the growth of any industry and compete globally.

EMPLOYEE PARTICIPATION IN MAKING DECISIONS

Once the employees are trained or qualified to handle the requirements of new technology or new processes, the supplier's leadership must create an environment for their active participation, innovation, and decision making. The leadership must create a system for empowerment or ownership by creating the management structure, clearly defining roles and responsibilities, establishing a system for accountability, and practicing good communication and reporting methods. Through effective communication, the employees must receive the objective feedback about their processes and must be allowed to perform as planned through a documented system.

Some of the characteristics of a successful leader are honesty, integrity, charisma, and future-oriented thinking. *Honesty* is the most critical. The leader must be honest with people and be willing to walk the talk. *Integrity* implies an ability to keep one's word. In other words, once the leader is committed to an action, that action must be completed as intended with the leader's active involvement. A *charismatic* leader inspires employees through personal beliefs and effective communication, by empathizing with their needs and recognizing good deeds. The *future-oriented* leader is always preparing the employees for the next steps, products, or services. This brings excitement to the team and uplifts team morale. The great leaders possess high self-esteem but low ego. They take the responsibility but rarely use their authority. The leader's true power comes from the *people*—not from the designation.

A leader ensures that people are performing to the best of their abilities. The leader must inspire their intellects, motivate their minds, and nurture their efforts. The leader inspires the intellect by setting up good values of mutual respect and commitment to meeting the customer's expectation and nothing less.

Challenging their mental capacity can only motivate the employees. The leader's goal must be to utilize employees' minds more than their bodies. The leader must empower people to set their own aggressive goals for improvement. This aggressive goal setting encourages employees to accept the challenge and enjoy the feeling of accomplishment. To nurture employees' effort, the leader must recognize small successes, encourage employees to learn new practices, reward the self-learning process, and challenge employees to improve. We all have needs that change over time. If employees have the opportunity and ability to extend their capabilities, they will continue to grow. The best thing a leader can do for a company is to set aggressive goals and give employees complete freedom to achieve those goals. Employees must be able to use their hands for effort, their heads for innovation, and their hearts for commitment in order to maximize their organizational and personal performance.

Employee involvement and employee development to utilize changing technology effectively are critical in staying competitive in almost any industry today. Involving employees in every possible level of decision making is the developing hallmark of superior companies. This means the alignment of job expectations with employees' physical, emotional, and intellectual abilities, without which motivation will be lacking and performance will suffer. When there is alignment between skills and abilities and job challenges, one can expect superior performance, noticeable accomplishments, and job satisfaction.

How can management enable this employee involvement? By treating employees as individuals, with unique identities and intellectual capabilities. Labor is more than the physical effort of employees. When they are treated this way, they can bring their intellectual inputs to bear on the betterment of the process—a knowledgeable workforce. Performance follows, problems begin to disappear, and employees raise the bar.

COMMUNICATION

There must be open communication between leadership and employees so that expectations and ideas can flow freely in both directions. The employees need to hear the expectations of the leadership, and the leadership needs to build credibility with the employees for working toward the common goals. Lack of this communication is usually an indication of other problems, quality issues being just one.

Employee involvement can be gained by establishing both formal and informal communication oriented toward quality. The formal communication establishes clear objectives and expectations, including rewards for achieving the challenging objectives. Such direct formal communication builds trust, in addition to demonstrating leadership's respect for employees and expressing its dependence on employees. Though mere open communication doesn't necessarily indicate the success of an enterprise, all successful businesses have such open communication channels.

WORK ENVIRONMENT

In addition to open communication, employee's education, skills, and experience need to be aligned with their tasks. This reduces stress for the employees and increases productivity, morale, and quality for the business. What follows then are customer satisfaction, business growth, and long-term success.

Attention to the work environment includes ergonomics, safety, work schedules, and appropriate tools for the tasks, allowing employees to utilize their full effort in maximizing gains for the business. All of these are the inherent goals of a Six Sigma initiative that aims to optimize the business for efficiency. Again, the emphasis is on doing the right things the right way, efficiently.

INTEGRATED THINKING

Since an organization is made up of management and employees, it goes without saying that if management believes in its skills and abilities, the same must be recognized in the employees. Taking an interest in employees' growth and development is a natural function of management. It takes a lot of effort to get talented and experienced employees, so their talent needs to be kept current and in step with changing technologies and processes. Employees with choices brought about by skills development work willingly, as opposed to the contrary situation, where for lack of choice work may be done unwillingly. Successful companies abound that take care of and take an interest in their employees, who in turn make committed efforts to make those companies successful. Paying attention to the needs of the employees is the first in a long series of steps leading to improvement, productivity, satisfied customers, and long-term success for the business.

Business organizations are living entities, with their functional bodies, and no part can be ignored, to the peril of the rest. Profitable companies understand this dependency and attend to the development of employees, recognize their contributions, and continuously identify the best performers to engage their intellect. While securing investments, environmental resources, and other valuable assets of society, if the leadership is not endeavoring to create value for the betterment of all involved, then its activity is no less than criminal. When the leadership treats the employees as though they are the customers, the employees then know how to treat *their* customers, ensuring customer satisfaction and the long-term success of the enterprise.

Precontrol Charts

Once the effort is made to enlist employees intellectually, the next step is to provide control over the process to the operator in a way that is simple to understand, interpret, and maintain. Precontrol charts are designed to give control over a process to the employees, who can then monitor the process and adjust it as necessary without going to a supervisor or engineer. The precontrol chart is a colorful visual tool for controlling the process. The precontrol chart is an extension of the control chart, but simplified to empower employees.

The precontrol chart is a tool for determining when the process may go out of statistical control and need adjustment before the out-of-control condition occurs. Precontrol is implemented at a later stage than the control chart. The control chart is used to maintain the process within statistical control. With the necessary adjustments made over time and the knowledge gained, a cookbook is prepared for process adjustments. Documenting process conditions and associated adjustments is critical in maintaining the precontrol chart.

The precontrol chart is simple to use, as there are fewer rules than for control charts, and there are fewer calculations, because it is a visual tool and uses a smaller sample size (2 samples). Precontrol rules are based on the normal distribution. Accordingly, the probability of having a data point in the green zone is 86 percent, and the probability for the yellow zone is 7 percent on each side. Figure 8.4 shows the precontrol chart. The chart has three zones, green, yellow, and red. The red zone represents the area outside the process control limits.

Setting Up the Precontrol Chart.   For effective use of the precontrol chart, the process capability should lie well between the specification limits. The process capability must be at least 75 percent of the tolerance ($C_p \leq 1.33$), and an even smaller percentage for tighter process control if the process is likely to drift. For setting up a precontrol chart to maintain Six Sigma capability, six times the standard deviation must be equal to or less than half of the tolerance.

To establish various zones, the tolerance is divided into four equal parts as zones, as illustrated. The zones are shown

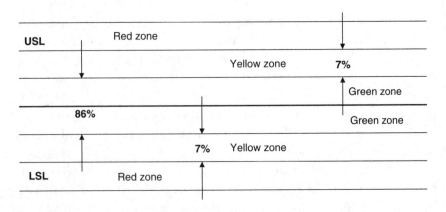

FIGURE 8.4.   Precontrol chart.

in colors—red is outside the tolerance limits, green represents the two inside quarters, and the yellow zones represent the outside quarters of the tolerance. To reduce the probability of going into the red zone, one can use control limits to define the four zones. Trusting that the control limits are within the specification limits, Fig. 8.5 shows the precontrol zones within the control and tolerance limits.

Before launching precontrol on a process, it is imperative that the process has been characterized and adjustments to be made to the process have been documented, capability has been established, and employees have been trained to use the precontrol charts. The frequency of measurement is established between the major process changes; for example, shift change at about 20 minutes.

Precontrol Rules.   The rules for using the precontrol chart are as follows:

- Setup rule: Production is set up to run if five consecutive parts show measurements in the green zone.
- However, when the process is running, it is required to take two samples and analyze them as follows:

  1. If the first sample is in the green zone, continue to run the process. There is no need to measure the second part.

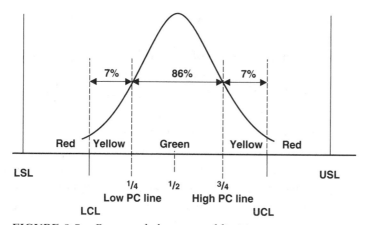

FIGURE 8.5.   Precontrol chart control limits.

2. If the first sample is in the yellow zone, check the second part. If the second part is in the green zone, continue to run the process. If the second sample is not in the yellow zone on the same side, adjust the process halfway from the last measurement, and continue to run the process.
3. If both samples are in the opposite yellow zones, shut down the process, correct the nonconforming condition, and start with the setup rule.
4. If the first sample is in the red zone, adjust the process, and start with the setup rule.

LEADING IMPROVEMENTS

Leading improvement is a process that must be adhered to if one is to achieve continual improvement and sustain the Six Sigma initiative. Processes to lead improvement have been identified in the ISO 9001: 2000 standards as well.

Management involvement and commitment are essential for the successful implementation of the Six Sigma initiative. The successful implementation of the Six Sigma initiative is the responsibility of the chief executive rather than that of the Six Sigma champion. Figure 8.6 shows the overall flow of this process.

MANAGEMENT REQUIREMENTS

Six Sigma objectives are defined and documented, and the necessary resources are identified and committed by top-level executives, before management takes up the commitment to implement Six Sigma. This commitment is demonstrated as the documented responsibility of the employees involved in the initiative. Resources needed to provide the education, training, and skills for employees are documented. The CEO appoints one of his or her direct reports to be responsible for leading the Six Sigma initiative and tracking specific goals and objectives. The leadership must also ensure that there is a process in place for data analysis, reporting, and remedial action to achieve the corporate goals and objectives.

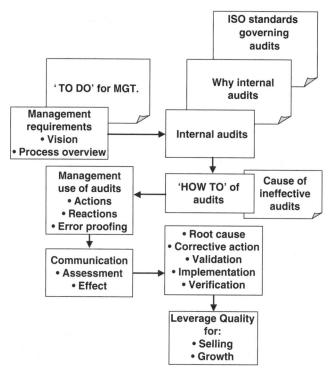

FIGURE 8.6. Flow of leading improvements for effective implementation of a process.

To-Dos for Management. The leadership's role is limited in implementing the Six Sigma initiative. The initiative suffers when management neglects the business aspects of Six Sigma and instead cares more for the feel-good side. This neglect may manifest itself in lack of commitment to the Six Sigma initiative, lower employee participation, and frustration in the Six Sigma community. This means that the Six Sigma initiative is just a responsibility of the Six Sigma Champion, who cannot be successful without the leader's open commitment and the employees' enthusiastic participation. The Six Sigma methodology and the experience of successful companies have proven that the CEO's understanding of the intent of the Six Sigma initiative and his or her passionate commitment are the highest-priority success factors.

Defining Employee Roles.   Clearly defined employee roles, responsibility, and authority are prerequisites for achieving dramatic improvement using the Six Sigma methodology. Many times, employees want to get involved; however, they are unable to prioritize their efforts toward Six Sigma. This occurs because employees do not see the indications of priority from their superiors, including the leadership. They see lack of consistency at the highest level; they see lack of interest at the middle management level; and they know that the Six Sigma initiative would not be successful without their support. Being smart, employees do not spend their resources on the Six Sigma initiative. If this misalignment between the published initiative and commitment through action is corrected, employees start holding up their end. The process starts with getting employees excited about participating in the Six Sigma initiative as a major corporate initiative, because even the executives are participating.

From this argument, the imperative is clear. The leadership must clearly establish corporate expectations for Six Sigma performance, for both management and employees. In other words, what would the corporation gain from Six Sigma? What would employees and management gain from the Six Sigma initiative? Defining the purpose of implementing Six Sigma, department managers' goals and roles, and employees' goals and roles in achieving corporate objectives up front is a necessity. Their contributions toward achieving Six Sigma objectives, corporate growth, and profitability must be recognized and publicized. Though achieving growth and profit are the ultimate goals, it must be clearly communicated that producing the highest-quality output in every process of the company, reducing waste, and finding innovative new solutions are of paramount significance.

AUDITING SIX SIGMA IMPLEMENTATION

To ensure that progress is being made and the Six Sigma initiative is showing results, an auditing team must be formed at the highest level to assist with compliance, promote the effectiveness of the Six Sigma initiative, and identify areas for improvement in the implementation of the methodology. This is not normally done, as Six Sigma methods are too new and impressive to be questioned. Given the level of resources committed to the initiative, the people involved, and the results expected, the complete cycle of implementation must be planned out. Normally, implementation is planned; however, verification is *not* planned, as it may point toward the corporate leaders, as they are most likely to be the cause of lack of progress.

Most companies have already implemented internal quality auditing functions. Financial audits are performed periodically as well. Similarly, auditors of the implementation of Six Sigma in each department should be independent of the areas they are auditing. The audits serve as the trigger for providing resources and necessary help in prioritizing activities. If there is a lack of progress in some areas, additional support could be provided. The objective in auditing is the integration of Six Sigma in daily activities.

Reasons for Audits.   Any process that is successfully implemented follows the PDCA cycle. One of the key elements of the PDCA cycle is the Check phase. In the Six Sigma PDCA cycle, certain resources are identified as vital to the success of the Six Sigma initiative. The Check phase of the cycle is the audit at the process level, and savings at the financial level. However, if one only looks at the savings without auditing the Six Sigma process, there is a good chance that the overall objectives of the Six Sigma initiative will not be achieved. This ensures the breakthrough performance of projects at critical processes in the business. The sustainable success of projects at critical processes is a must for the realization of the Six Sigma vision in the company; otherwise, the Six Sigma initiative could become a fiasco.

The implementation of Six Sigma audits requires an acknowledgement and understanding on the part of management that many company weaknesses will become visible and will subsequently have to be fixed. Good Six Sigma process will produce more good breakthrough solutions; fixing process weaknesses must become a high priority. Only audit findings will give the management confidence that the company will be able to achieve the planned benefits of the Six Sigma process. One of the major problems with the Six Sigma initiative occurs when the leadership is not involved and results are not realized as planned. Even worse, expectations for the Six Sigma initiative are not even planned sometimes. The only way to prevent failures and the frustration of the Six Sigma initiative is to enforce auditing of the Six Sigma process.

Planning for Audits.   Auditing the Six Sigma initiative must be a planned activity. The audits must be planned while developing the strategy to implement Six Sigma. Audit planning addresses issues such as the importance of the project based on the criticality of the associated process to customer feedback, the internal need for process improvement based on waste of resources, or the defect rate. The planners may also consider using the existing pool of qualified internal quality auditors. Auditing the process must not be considered as a threat to Six Sigma in the planning stage; rather, it is a necessary step to ensure that the investment in Six Sigma is not wasted. Audits must incorporate leadership, strategy, planning, training effectiveness, project selection, results, and sustainability.

Six Sigma audits should be performed by someone who is knowledgeable, trained, and significantly experienced (at least 10 years is recommended) to ensure the integrity and independence of the audits. One can use external auditors because internal auditors may have difficulty in questioning people in management or the leadership positions. In either case, standard procedures for internal audits must be followed.

Scheduling Audits.   The auditing schedule must consider the urgency of implementing Six Sigma and achieving results. If the corporation is planning to implement Six Sigma because of external drivers, the Six Sigma initiative may conflict with internal priorities. In this case, the audit frequency could be quarterly or semiannually. If the drivers for Six Sigma are aligned with internal needs, the priorities will not be in conflict and minds will be aligned. More people will be involved and progress will be made more rapidly. In this case, monthly audits of some aspects of the Six Sigma initiative would be helpful in maintaining awareness of the urgency of implementing sound processes. Unlike quality system audits, the Six Sigma audits must be planned on a quarterly basis due to the changing requirements and the progress of the initiative in the company.

Conducting Audits.   Since Six Sigma is a highly visible, high-value, and high-energy initiative, one must use a prepared checklist as a guideline to avoid emotional confrontations in conducting the audit. A sample checklist is shown in Fig. 8.7. In preparing for the audit, auditors should review the Six Sigma planning documents to understand the intent and objectives of the initiative at the corporate and department levels. The intent of the audits must be to assess the sanity of the approach, the use of the methodology, and the results and savings achieved, and to identify opportunities for improvement. A day prior to the audit, the auditor should contact the departments to be audited to ensure that the time is convenient for them and that people are available to interact with the auditor.

During the audit, the auditor must be able to express curiosity, similar to that of a 10-year-old child, instead of pretending to be an expert in Six Sigma. The auditor must suppress the urge to make recommendations for a specific project while conducting the audit. The auditor must not take any response for granted and must be prepared to challenge the status quo. Internal audits are conducted in a friendly team environment. The Six Sigma audits must be viewed as a joint exploration by the auditors and the auditiees. The Six Sigma audit is a critique of a process by an auditor or team of auditors, with the intention of helping the auditee or the Six Sigma initiative. Auditor and auditee must look for disconnection, lack of progress, bottlenecks, opportunities for acceleration, ways to reduce cost (effort) or cycle time (speed), and evidence of alignment with corporate objectives.

Audit Report.   Once the audit is complete and opportunities for improvement have been identified, the report does the rest.

PROCESS	SIG SIGMA AUDIT QUESTIONS	ASSESSMENT
Commit to Six Sigma.	Leadership has been trained in Six Sigma concepts and methodology.	
	Leadership has identified drivers for implementing Six Sigma.	
	Business opportunity analysis has been performed based on market position, competitive assessment, and corporate performance.	
	The cost and benefit analysis of the Six Sigma initiative has been completed and utilized in making the commitment to Six Sigma.	
	Executive compensation or incentives have been linked to success of the Six Sigma initiative.	
	A commitment to Six Sigma and expectations in growth and profitability have been developed and communicated to all employees.	
	The CEO has established a quantifiable and verifiable measure of success of the Six Sigma initiative.	
Appoint a Six Sigma corporate leader.	The corporate Six Sigma Champion directly reports to the CEO or equivalent.	
	The Six Sigma Champion has adequate education, sufficient analytical skills, and strong communication and presentation skills.	
	The Six Sigma Champion possesses necessary leadership, sales, and persuasion skills.	
	The Six Sigma Champion has been given a very clear job description identifying goals and objectives, expected deliverables and performance measures, and potential compensation.	
	The Six Sigma Champion has the necessary resources and authority to make decisions about the Six Sigma initiative.	

FIGURE 8.7.   Six Sigma Audit Checklist.

Process	Sig Sigma Audit Questions	Assessment
Identify key areas for profitability improvement.	A list of potential projects based on the business opportunity analysis that would contribute to the profitability of the company has been developed.	
	Projects have been prioritized based on their impact on customer satisfaction, performance improvement, sales, profitability, or employee satisfaction.	
Planning for Six Sigma	The CEO's staff has been trained in project management, statistical thinking, innovative thinking, and process thinking.	
	A Six Sigma roadmap and an implementation plan have been developed with clear milestones in various departments.	
	Critical resource requirements for making the Six Sigma initiative successful have been identified.	
	The corporate organization chart includes the responsibility of Six Sigma initiative, and other key players.	
	Departmental expectations have been established, documented, and communicated.	
	Barriers to the progress of Six Sigma have been identified and addressed for mitigation.	
Implement performance measurement system for the Six Sigma initiative.	A corporate scorecard has been established to monitor savings and changes in growth and profitability.	
	Periodic communication meetings to report progress of the Six Sigma initiative have been scheduled.	
	Performance levels and trends of key performance metrics (Q,T,C) of each department are reported.	
	Executive review process of the Six Sigma initiative has been established.	
	Method to share company's success stories with stakeholders has been set up.	
	Action items to remedy the bottlenecks and correctives actions are publicized.	

FIGURE 8.7.   Six Sigma Audit Checklist. (*Continued*)

PROCESS	SIG SIGMA AUDIT QUESTIONS	ASSESSMENT
Establish a Six Sigma performance-driven compensation system.	A process for CEO recognition of employees for exceptional improvement has been established and implemented to inspire more success stories. A plan to share savings from the Six Sigma projects has been established and communicated to all employees. A performance review system has been established to assess employee performance against the goals and objectives.	
Conduct basic training for employees.	Process and statistical thinking training is provided to all employees, including the leadership. Overview of the Six Sigma methodology has been presented to and understood by all employees. Innovative thinking training is provided to all employees for achieving breakthrough process improvement. Training for management time is offered to all employees for managing multiple projects or priorities. Corporate vision and values are communicated to, understood by all employees, and used in making decisions.	
Conduct project-driven Six Sigma training.	Six Sigma training needs have been assessed based on the list of projects. Candidates for Black Belt and Green Belt training have been identified based on established requirements. Green Belt and Black Belt training schedule has been established along with the projects. Expectations for improvement for Black Belt and Green Belt certified employees are clearly communicated.	
Verify quantifiable impact on bottom line because of above projects.	Project completion is verified against established goals.	

FIGURE 8.7.   Six Sigma Audit Checklist. (*Continued*)

PROCESS	SIG SIGMA AUDIT QUESTIONS	ASSESSMENT
Verify quantifiable impact on bottom line because of above projects.	The accounting department verifies savings due to the process improvement for each project. Corporate scorecard is updated, assessed against the profitability and growth, and reported to employees.	
Recognize success.	An employee recognition program for successful completion of projects has been established. Employees are encouraged for innovative solutions through incentives. Successful completion of projects is documented by employees and project leaders. There is a platform or framework for employees and project leaders to share their success stores and lessons learned. A project depository has been set up to archive process improvement results and knowledge gained.	
Develop internal expertise and resources.	New candidates for Black Belts and Green Belts are periodically reviewed and trained. Experienced Black Belts and Green Belts are actively involved in helping other teams when needed.	
Institutionalize Six Sigma methodology.	A process to identify projects for continual breakthrough improvement has been established. New teams are formed to work on projects regularly.	
Monitor and sustain improvement.	CEO and staff review the Six Sigma initiative regularly for its adequacy and suitability. Executive team updates the Six Sigma initiatives with new ideas for maintaining employee excitement and commitment. Employee feedback is sought and incorporated to improve the Six Sigma initiative.	

FIGURE 8.7. Six Sigma Audit Checklist. (*Continued*)

The report must be prepared to get management's attention. It must highlight the proper issues to ensure that the audit's results are not overlooked. The report must create a sense of urgency to remedy any bottlenecks. The report must be concise and direct to convey the need for any intended corrective actions. The company's management must ensure that all employees understand the corporate commitment to Six Sigma for its growth and profitability benefits. They should allay any employee fears for genuine lack of progress, and encourage employees to continue to achieve innovative solutions through recognition for the successful completion of projects.

Caution against False Audits. One must take precautions against audits that report everything to be good. If the audits look too good, they must be audited, too. It is only human nature to avoid reporting lack of progress for fear of reprisal. This is why the independence of auditors and the commitment to make the company successful must be overriding objectives in capturing and reporting audit findings. Employees must be assured of support in the event of failures and celebrations in the event of successes. For example, if one makes a major mistake that hinders the project's progress, but the mistake is a learning experience, the leadership must understand that. However, repeated mistakes must not be allowed, as they represent carelessness. The Six Sigma audits must be considered the management's eyes and ears for gathering performance information.

Using Audit Findings. The Six Sigma Champion has a significant role to play in making the initiative successful by utilizing the audit findings as an effective management tool for the company. The Six Sigma Champion must first understand the value of the audits, then persuade the executive management.

Today, many large corporations have several thousand Six Sigma projects. Six Sigma audits would help to identify areas of insufficient improvement and raise warning flags that additional support is needed for those areas. The audits could also be used to get management's attention regarding its lack of active participation. For example, say that it appears that the CEO of a large company is not actively involved. Six Sigma initiatives in that company may not produce glaring successes. To gain the CEO's buy in, the audit checklist can be used as a great follow-up tool.

Six Sigma audits work like preventive medicine. They help to identify problems in the Six Sigma initiative while they are small, when there are more opportunities to correct them, and the cost of changing them is small, too. Without the Six Sigma audits, the CEO will depend on the financial results. Sometimes discovering problems through the financials could become a major disaster.

Correcting Audit Findings. One's reaction to challenges to the Six Sigma initiative can be twofold. One approach is corrective action to remedy an unacceptable condition, so that the symptoms that indicated the problem are eliminated. This may or may not cure the problem. In fact, this approach may create problems elsewhere in the business process. The second response is along the lines of institutionalizing the solution. This is based on the root-cause analysis at the company level, linking all the vital functions of the business, including the leadership function. This overview provides the opportunity for improvement that can be addressed through innovative and systemic actions. In both these cases, the focus needs to be on robust methods that can errorproof the process and decrease the dependence on individuals. In either of the cases, the following steps are necessary.

Assessment of the Problem's Effect. Since there is much contention for the company's limited resources, it is important to convey the severity and significance of a problem. The magnitude provides the significance, and the severity shows the ratio of the problem to the costs directly and the profits indirectly. Remember that even a minor problem can affect profits adversely. Such analysis sensitizes management to the critical nature of the information being provided. Instead of reporting the effect as a small fraction of the total activity, it can be reported as a percentage of total profits, which is naturally a much larger number, providing the necessary alarm. If the problem reduces the profit of a product line by 80 percent, that is quite significant. The same effect on the total enterprise might be 0.5 percent.

Root-Cause Analysis.   The art and science of root-cause analysis is itself a vast discipline. However, the basic approach from a management perspective is the same in all cases. In the PDCA cycle, this analysis falls under the Act phase. At this stage, one assumes that the root cause must be identified. Using cause-and-effect analysis, a cross-functional team can narrow the possible causes to a few candidates, and with further analysis of the available data, pinpoint the root cause. The cause may lie in one of four major areas: materials, machinery, methods, and people. In analyzing the root causes of problems, one should look at the definition of the problem, check for the correct use of tools, and look into lack of resources.

Corrective Action.   The task of remedying the Six Sigma project problem must assigned to the project leader. There are many approaches to corrective action, depending on resources, urgency, and long-term technology plans. The appropriate corrective action must be addressed at the system level, instead of just at the individual project level. The corrective action often leads to the office of the CEO or the staff. The Six Sigma Champion must be able to establish a clear action item for the CEO's office. Otherwise, the Six Sigma initiative starts to fall apart.

CORRECTIVE ERRORPROOFING

When developing a corrective action to remedy a deficiency in the implementation of Six Sigma, one must attempt to do errorproofing, which starts with the identification of the opportunity for improvement. To errorproof the implementation, one must look into failure-free methodologies—that is, the ones that will work regardless of circumstances. The use of software tools could be a good way to prevent mistakes due to programming flexibility.

Before implementing corrective action at the project level, one must validate the solution. Implementing the wrong solution wastes the company's valuable and limited resources. The project leader must ensure that appropriate communication and training are maintained throughout the change process. This includes verification of the understanding of the operator of the data collection results, and, above all, that the changed process is producing the desired results.

MANAGEMENT REVIEW

The most significant factor in sustaining the Six Sigma initiative is the executive review of the results. Operations or financial reviews are conducted monthly. At the same time, the initiative can be reviewed for progress and its effect on the bottom line. The review must follow the standard management review procedures for scheduling, attendance, agenda, and minutes of the meeting.

The Six Sigma initiative needs an effective executive review to prevent its deprioritization. This means the right frequency, agenda, action items, attendees, and intent. Lack of any one of these can lead to a fruitless management review. Done correctly, these reviews can lead to improvement in the sigma level and benefit the company.

Management can use internal audits as its eyes and ears, corrective and preventive actions as its arms and legs, and the management review as its head and heart in providing direction. The audits to find nonconformance, corrective actions to remedy faults, and management review to ensure that there is harmony in the three-pronged process all lead to the betterment of the quality system and the business organization. Since quality is the responsibility of all members of an organization, every process is included in the quality system as a powerful business management tool, which leads management to take the responsibility for ensuring its definition, documentation, and implementation.

SELLING SIX SIGMA SUCCESSES

Having implemented Six Sigma successfully, the executive management must become the best salespeople for the company. It is critical that the CEO makes the successes visible inside and outside the company. Project leaders must be encouraged to network with their peers inside and outside the company and share success stories. After achieving savings and improving profitability, telling the world that you have a best-in-class system is a great way to sell more products and services. Therefore, the company's accomplishments must be publicized through newsletters, websites, conferences, articles, or forums.

COUNTING THE CHANGE

In the absence of good business performance measurements, businesses count on the financials to gage the health of the company. With all the competitive pressures and shrinking margins, businesses are practically left with counting the change. That's their profit. One of the key steps in Six Sigma efforts is measuring the effect of the change produced by the improvement efforts. Of all the elements of Six Sigma efforts, this one is much more open to error and lends itself to the delusional efforts of management under pressure to make the numbers. Because of this, one must take care in how one counts the change.

Knowing that Six Sigma is a fact-based improvement methodology, one must measure the status of the initiative. It is good to remember that what is measured is delivered. Measurements must also adapt to changing business conditions. Those who know the rules of measurement will succeed, and others may only be playing the game without knowing the rules. Caution is the watchword when someone says that they have the playbook, and everyone needs to play by it. You may well predict what the results will be under these conditions.

Just as an individual's health is checked, measured, and tracked, a business needs to monitor its health by measuring its performance against its goals to identify areas for improvement. It used to be simple to measure the productivity of a business and equate that to performance in good times. But technological innovations are moving toward complex product functionality that provides customer appeal, and measurements must reflect the customer-related realities.

Technology is compressing time and space, and a global marketplace is shaping up, with global competition and suppliers. Because of these new dynamics, there is an increasing emphasis on performance and innovation to increase, maintain, or shore up declining margins. If one adds mergers and acquisitions and the forces of convergence to this mix, a need for a comprehensive measuring and scorekeeping system emerges.

These forces have led to many different scorecards. The ISO standards, Malcolm Baldrige National Quality Awards (MBNQAs), and Six Sigma success measurement methods led to corporate performance measurements. The spiral of demand and supply changed the focus of business from quantity to quality and functionality.

J. M. Juran's Financial and Quality Trilogies in *Juran's Quality Handbook* (1998) showed the relationships among planning, control, and improvement in the financial, quality, and business spheres. They all have inherent limitations, as they do not include all employees who are in the best position for internal improvement.

The leadership that creates the vision lives in the future, the management likes the known certainty of yesterday, and the employees face the reality of now and today. Each one sees the other two with some distain and distrust because of past behaviors, all stemming from poor measurement methods. Ideally, the leadership creates goals for a superior, successful, and profitable organization. The management team manages and implements dramatic improvements for higher quality in the shortest possible time. Employees innovate product and service improvements and deliver the same to customers.

Standards and Scorecards

The ISO 9000 standard is a business management system that provides the impetus to move business performance responsibility to all departments. The newest version of ISO 9000 requires measurements of the effectiveness of key processes. The ISO 9001 requirements follow a PDCA process:

- Plan for set up
- Do for excellence
- Check for verification
- Act for improvements

The MBNQAs were established in 1987. They provide the Baldrige Criteria, a system for managing process performance. These criteria are based on the best practices in the following areas:

- Vision
- Customer focus
- Employee involvement
- Process excellence
- Market leadership
- Superior financial results

The Balanced Scorecard, developed by Kaplan and Norton (1996), provides a strategic management system using four business perspectives (Fig. 8.8):

- Financial
- Customer
- Learning and growth
- Internal business processes

For each of these four areas, the enterprise examines the following factors:

- Goals (How will you define success?)
- Measures (What is the measurement to show success?)
- Targets (What are the qualitative objectives?)
- Initiatives (What activities will you launch to achieve them?)

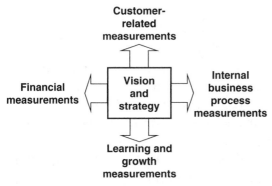

FIGURE 8.8. Elements of the Balanced Scorecard system. (*Based on Kaplan and Norton, 1996.*)

The Balanced Scorecard reveals much more than the traditional financial view. It also links the objectives of each of the four perspectives. Each organization selects specific measures and draws links between them. This necessarily top-down process enables work groups to devise their own business scorecards to show their contributions to the strategy of the organization. In implementing the Balanced Scorecard, managers outline their strategy, and departments go through training and develop the vision, strategy, and measurements. They develop goals and objective targets. They identify weaknesses and construct action plans. Though the Balanced Scorecard has been proven successful in the banking, oil, insurance, and retailing industries, the practicality of its implementation at the business grassroots level has been questioned.

With the ever-increasing implementation of Six Sigma methodology, there arises a need for a new performance measurement system that can relate the operations performance to the financials. With accelerating trends in technology, high expectations for performance, eroding prices, and shrinking margins, business needs a performance measure that addresses all aspects of the business in a robust and integrated manner.

Business Processes.   A business is a collection of processes, such as sales, purchasing, marketing, operations, quality control, and engineering. The process inputs for a business are capital, material, information, and people. The outputs of any business are products and services and associated variances, variance being the cause of excessive cost and loss of profits.

Any attempt to measure performance needs to measure the variances in order to reverse, control, or eliminate excessive variance. The Six Sigma Business Scorecard is designed to measure all aspects of an organization. It addresses the concerns expressed by executives about existing scorecards, such as their irrelevance to the employees who do the work of creating changes. Scorecards designed with mostly strategic intent do not flow down to the process measures needed for the strategy to succeed. Figure 8.9 shows various elements of the Six Sigma Business Scorecard, which promotes inspiration by leadership, improvement by managers, and innovation by employees.

Need for the Six Sigma Business Scorecard.   A survey of a cross-section of employees indicates that they have no idea how leadership plans to improve profitability. People know what they are supposed to do, but only about a third of them know how well they do. An even smaller percentage of employ-

FIGURE 8.9.   Elements of the Six Sigma Business Scorecard.

ees know the rate at which their business processes have been improving, especially nonproduction processes. This implies a continued focus on delivery, instead of getting better and knowing the process for improvement. Generally, employees are closest to the customers and most familiar with their needs. Though employees know of measurements for the short-term success of their business, they do not have measures of performance for various aspects of their jobs and their relationship to the long-term success of the company. Employees may feel good about their products and services, but are uninformed regarding the performance levels expected or the rates of improvement targeted.

Insufficient Process Performance Measurements. Businesses utilize many financial measurements, such as sales and productivity, margins, assets, and profit, but they lack good measurements for operational effectiveness, which in reality drives the financials. Companies in niche markets with light competition may enjoy good margins and be able to tolerate some waste without operational measures. However, companies with stiff competition, complex products, and complex systems will have a tough time without operational measures. The increasingly complex businesses with several products and services, many of them integrated and multilocal (global locations), face a greater challenge in their measurement efforts.

In the quote-to-cash process, not everything moves according to process plans or expectations. The salesperson's focus is different from that of the design team, which again is different from the production team's needs. If you throw in the unexpected variations of customers going bankrupt and suppliers failing to meet requirements, the challenge intensifies. The bottom line is that some targets, such as delivery dates, are met, but the customer complaints, returns, and rework highlight the reality of the need for better controls and measurements. Some companies do have performance measures that create reports and charts. Other businesses may have many measurements, but not the right ones. Some business units are imposed with dictated measurement from corporate headquarters, leading to ineffective communication at all levels of the company. The net result is that the measurements many businesses deploy today are ineffective, as they are not visibly linked to growth and profitability. Insufficient or ineffective measurements create corporate-level inconsistency in decisions and performance.

Measurement in Quality Systems. The process model shows activities utilizing resources to transform inputs to outputs. Each process needs control of its inputs, in-process steps, and outputs. To ensure compliance, verification methods and metrics are needed at all the process stages, from supply through delivery. The process model works when the process owner ensures that the following requirements are met:

• Clear documentation
• Trained employees
• Appropriate data collection
• Data analysis
• Corrective actions

Process management requires that employees, including management and leadership, understand a set of defined and documented roles and responsibilities, and are held accountable for their performance. Process management implies a focus on the business process and good teamwork due to clearly defined roles and responsibilities. The benefits of process management include the following:

• Process consistency, standardization, and repeatability
• Customer confidence
• Increased business
• Happy leadership

The Six Sigma initiative takes the business processes to the next level, beyond process management. It challenges the process owners to improve the performance of a well-defined process through the following methods:

- Process thinking
- Accountability
- Aggressive improvement goals
- Communication
- Innovation
- Metrics
- Rewarding experience

Typical measurements that aid in process monitoring and management include the following:

- DPU
- DPMO
- Process yield
- Customer satisfaction
- Customer returns
- Employee suggestions

Executed well, when the Six Sigma methodology achieves improvement objectives, and the Six Sigma Business Scorecard helps businesses to achieve their profitability objectives. The Six Sigma Business Scorecard reflects the effectiveness of all processes at the action level, rather than the strategic level. This scorecard relates to the entire organization, instead of viewing the organization from the measurement perspective alone.

SIX SIGMA BUSINESS SCORECARD

Profitability is a business target but not the main purpose of business, which is to provide products and services that customers want, leading to profits. If a business uses up society's valuable resources, burdens the environment, and is not profitable, it is wasteful and unjustified. A business's objectives are to provide products and services to customers through its core competency, care for its customers, develop its people, and be profitable to shareholders.

A sales- and revenue-growth focus leads to suboptimal processes without correlation to their contribution to the planned revenue growth. Just looking at the current accounting systems, with their emphasis on the detailed tracking of costs, one can see lost opportunities, hidden costs unaccounted for, and vast opportunities for improvement. In simpler business models, accounting plays a main role in providing decision support information. In complex business models, decision support data necessarily needs to come from performance measures.

Management more frequently reviews profitability than the input streams that generate the profits. Another weakness is the lack of contingency planning for uncontrollable changes in external factors. Also, the leadership role needs to change from the oft-demonstrated self-interest and excess of recent times. A good performance measurement system considers not only external factors but also the role of leadership. A good key principle in the design of measurement systems is that *what is valued must be measured*. A good system meets the following criteria:

- Strategic in intent
- Clearly linked to organization's processes
- Easy and inexpensive to implement
- Easy to understand and communicate
- Includes analysis and reporting
- Includes improvement mechanisms

The Six Sigma Business Scorecard is a complete corporate performance system that requires leadership to inspire, managers to improve, and employees to innovate to achieve the optimum levels of profitability and perpetual growth.

This approach incorporates proven practices in process measurements, regardless of responsibility and authority. It holds the leadership accountable for business success through its active involvement. The Six Sigma Business Scorecard positions itself as the next step in the evolution of performance measurement systems. It is versatile in all business, service, public, and governmental organizations. It allows for total leadership and pinpoints improvements on an ongoing basis. The Six Sigma Business Scorecard combines 10 corporate-level measurements into seven elements:

1. Leadership and profitability
2. Management and improvement
3. Employees and innovation
4. Purchasing and supplier management
5. Operational execution
6. Sales and distribution
7. Service and growth

These are not necessarily universal but are designed to assess the effectiveness of critical activities that lead to wellness and profitability. Customization to the specific organization is effective rather than exceptional.

Leadership and Profitability.   As goes the leader, so goes the organization. Leaders are recognized by their honesty, knowledge, integrity, and visionary outlook. Their business activities become an extension of their lives, with the purpose of providing service to society, value to customers, growth for their employees, and enrichment for themselves. Their activities and behaviors need to match the values they set for the company, being the change they want to see. Any gap will be readily noticed by employees and copied to the detriment of the company. Since leaders are also lifelong learners, the performance measurement system needs to promote such openness and growth in the following leadership processes:

• Establishing, communicating, and practicing corporate values and beliefs

- Facilitating a strategic plan for the company
- Injecting positive energy into the organization
- Being informed in real time through performance measures
- Committing to the well-being of employees, customers, and society
- Guiding the staff to achieve the planned performance levels
- Challenging everyone to excel and innovate

Management and Improvement.   Middle managers are a link between strategy and results. They are caught between conflicts every day, and the tendency to meet the numbers of management by objectives leads them astray from quality or process goals. If management by objectives is changed to management by *process* objectives, there is a chance to move to statistical thinking, leading to fulfillment of strategic goals. Poor results indicate a need to adjust the process, preferably through self-correction. Management needs to relate process performance to profitability, since even a 1 percent waste in operations can lead to a double-digit percentage loss in profitability. Therefore, the focus needs to shift to the rate of improvement in operations. An improvement of 10 percent or less can be lost in accounting or measurement errors. There is a need for a higher rate of improvement, of about 30 to 60 percent, that can challenge employees to innovate. The improvement is visible and benefits all employees.

Employees and Innovation.   As more productivity tools are employed, the focus needs to move from managers to workers, to challenge and utilize their intellect. When continued learning and innovation take place at the worker level, a sense of accomplishment and job satisfaction results. This requires effective job planning, incentives for improvement, recognition for problem prevention, and rewards for innovation. Successful companies encourage the intellectual output of all employees, inside and outside the workplace, to bring out their best. To meet the customer requirements of better, faster, and cheaper solutions, the intellectual involvement of all employees is elementary. Such learning organizations will have the competitive edge in a global marketplace.

Purchasing and Supplier Management.   In business, about a third of costs are due to the cost of material or supplies. Therefore, creating value through this process has become an explicit necessity beyond just finding the cheapest source. Since a chain is only as strong as its weakest link, optimizing this process includes other defined requirements and goals and an aggressive rate of improvement. With increased dependence on suppliers, customer requirements for better, faster, and cheaper performance must be passed on to the suppliers.

Operational Execution.   Many exceptional strategies fail due to poor execution. Simpler and well-known principles, when well executed, lead to performance and profitability. Good operational execution requires passionate leadership that demonstrates personal involvement for the corporate good. Even Six Sigma initiatives have failed to yield results due to the lack of leadership involvement in execution. Another component of execution is *empowerment*, including education, authority, accountability, and recognition. There must be an equal emphasis on recognizing success and rewarding it, as well as on recognizing failures and the lessons learned from those experiences.

Sales and Distribution.   As the dependence on suppliers increases, the risk of losing a big chunk of sales increases, too. Risks to business viability as well as the risk of supplier viability increase with close relationships. For a business to grow continually, the dependence on one or two customers must be reduced, and new customers must be sought aggressively. The objective is to acquire a certain level of new sales. The challenge is that when a few customers provide a lot of business, the company becomes profitably complacent, which causes anxiety due to variations in economic conditions, or due to acquisitions and mergers. The leadership must set some goals to diversify the customer base and acquire new customers, regardless of the sales level at a given time.

Service and Growth.   The business growth aside from new customers can come a little easier by serving current customers better and strengthening the bond between the customer and the supplier. Besides, the growth comes from innovation and the introduction of new products or services. New product development must become part of the growth strategy and must be managed accordingly. In the case of service, the ultimate measurement is customer satisfaction, which is never fully achieved due to the changing customer requirements. The maximum score a corporation gives itself for customer satisfaction is 90 percent. So there is always room for improvement in customer satisfaction, regardless of what customers say. Meeting customer requirements and anticipating their expectations shows care on the part of suppliers, leading to superior customer service and customer loyalty.

Business Performance Index.   One dilemma for the CEO is what to look for to predict the corporation's financial performance. There are so many numbers, so many products or services, that it is a challenge to make sense out of measurements other than the profit and loss. This is a big handicap, because the CEO cannot grab hold of something, some process or product, and take care of it. The Six Sigma Business Scorecard includes a business performance index (BPIn) that is based on 10 critical measurements for establishing the wellness level of a company. The BPIn is the sum of weighted corporate performance measures in various categories of the Six Sigma Business Scorecard, as follows:

Measurement	*Significance*
Employees recognized for excellence	15
Profitability	15
Rate of improvement, all departments	20
Recommendation per employee	10
Total spending/sales	5
Suppliers' defect rate σ	5
Operational cycle time variance	5
Process defect rate σ	5

New business, $/total sales $	10
Customer satisfaction	10
Total	100

Performance is measured against the planned corporate growth and profitability goals, and the index contribution by each category is determined. The sum of all weighted scores for each measurement determines the BPIn. The typical performance index for a company that has not initiated the Six Sigma initiative or accelerated the rate of improvement, hovers around 55 percent.

Corporate Sigma Level. Six Sigma is a measurements-based, fast-improvement-driven, very focused methodology. The objective is to achieve lots of improvement very fast. It requires changes in the way of doing business. Significant energy and resources are committed to achieve lots of improvement. Therefore, a measure of its own success must be established at the corporate level. The CEO and the staff are at the helm of the Six Sigma initiative. For the success or failure of the initiative, only the CEO and the staff are responsible. In Six Sigma terminology, they are the opportunities for error that must be accounted for. According to Gupta's model, the corporate sigma level is based on the BPIn. The method of determining the corporate sigma level is defined as follows:

Corporate wellness (the measure of goodness)	BPIn
Corporate DPU	$-\ln(BPIn/100)$
Corporate DPMO (based on no. of executives)	DPU × (1,000,000/no. of executives)

The corporate sigma level is determined using the preceding DPMO.

For example:

| BPIn | 72.5 percent |
| Number of key decision-making executives | 9 |

Corporate DPU	0.322
Corporate DPMO	35,731
Corporate sigma	~3.3

Many companies are having difficulty in sustaining their Six Sigma initiatives. It is understood that if one values something, it should be measured. Since corporations value Six Sigma, they must measure the sigma level of the corporation and set goals and plans to achieve the desired sigma-level performance. The Six Sigma Business Scorecard requires that organizations understand the BPIn measurements in the process category and review the operational effectiveness in that context. Linking the sigma level to profitability is the fundamental reason for the success of the Six Sigma methodology.

If the sigma level is not acceptable, or the initiative is not progressing as planned, various elements lead to remedial action at the corporate or departmental level. To sustain the Six Sigma initiative, each department must contribute toward achieving the corporate sigma level and the corporate objectives.

The following are the steps in implementing the Six Sigma Business Scorecard:

- Creating awareness
- Building the business model
- Establishing the BPIn
- Establishing the Six Sigma Business Scorecard measurements
- Ensuring data collection capability
- Outlining a war on waste
- Managing change
- Integrating technology and the Six Sigma Business Scorecard

A common theme throughout our discussion is that an inspiring leadership team can bring about desired changes in the business to reach its goals. Learning from the first implementation of Six Sigma at Motorola from 1987 to 1992, Chairman Robert

Galvin provided the role model for leadership and behavior. During that time the quality vision, employee involvement, innovation, recognition, customer focus, and profitability served as examples that motivated other leaders and business to practice Six Sigma to achieve new levels of performance.

KNOWLEDGE MANAGEMENT

In all human endeavors, *laws* are broken now and then, and sometimes without consequences. But when the same activities break a known (or unknown) *principle,* the consequences are unavoidable and certain over time. The challenge is to know the obvious and subtle principles that apply to one's work, and to optimize one's work within those constraints.

Business wrecks have abounded in the past few years. Which principles have been violated? Some are obvious; others are not so obvious until the results show up in the courts of inquiry.

PROBLEM OF CONSENSUS

If you are working in a team environment to create the next thoroughbred to give your company a chance at winning, you cannot rely on consensus alone to make the horse happen. Consensus can create a camel—very good on its own accord and for its own application—but not for a horse race. A thoroughbred raised by a dedicated and dreamy horsebreeder has the potential to win a horse race! The principle here is that *majority is neither truth nor justice.*

The challenge to leadership in business is to find the working principles and align the company's business activities with them. Consensus has its place in political dealings, but agreements need to align with principles to be effective and *long lasting.* When you move into technical fields, the acuity of the principles involved becomes greater, and consensus needs to take a back seat.

CHANGES IN WASTE AND INEFFICIENCY

One emerging principle that needs to be heeded is the concept of knowledge management. Natural resources, raw materials, transportation, manufacturing, communication, and information management have progressed over the years to their levels of maturity today. Underutilization, waste, inefficiency, and poor management were tolerable in some of these areas, and the consequences were not life threatening to the organization or its natural and social environment. However, the speed of change in management tools based on knowledge and global competition now leave very little room for error, much less for deliberate mismanagement.

DATA, INFORMATION, AND KNOWLEDGE

Information management itself was born of data processing management information systems (MIS) and information technologies, and progressed to a C-level office, the chief information officer (CIO). Recently there has been an exploration for a "chief ignorance officer," one who will keep track of what the corporation *doesn't* know. Never let the irony be lost that the more you know, the more you will *not* know.

But information management is in transition, and it is showing the same strains that data overload created not long ago. If you replace *data* with *information* in the preceding statement, it depicts the situation today, drowning in information, needing to distill it into usable knowledge.

One can either agree that "We cannot let the monkeys run the zoo," as the CEO of a now-defunct corporation once described employee involvement, or work to create an "information democracy," as in some successful Japanese companies (Kao) that model a knowledge-creating and knowledge management environment where every member of the company has access to most company information except personal records. Those who contribute can be from any part of the organization.

FOCUS ON RESULTS AND SOURCES

Corporations know where their patents, important legal documents, and trade secrets are, just as farmers know where their eggs and fruits are. But farmers also know their own chickens and fruit trees and the ideal environment in which they produce and are productive. Farmers also keep an eye on developing technologies and methodologies to improve the business. When you apply this analogy to the organization, you see some important gaps.

Knowledge creation is one of the primary functions of a corporation, no matter what it produces. Just listen to the team members next time you have a team meeting. All the questions thirst after *knowledge* of what, where, when, how, who, and why. The *who* is deliberately put toward the end, because the old cliché, "It's not *what* you know, it's *who* you know that counts" plays havoc with organizational effectiveness. The cynical even go so far as to say that it's what you know *about* who that counts. This may sound as though all information is treated as knowledge, but it does bring the saying "Knowledge is power," to examination in a new light.

INFORMATION TO KNOWLEDGE

The information age was ushered in by the combination of computing power and communication networks. Today we certainly have more information than we need, and we are drowning in information—*information overload* is a common phrase now. The natural step is to question the information focus, though we still need and use information, just as we still need and use data, and evolve to incorporate the knowledge that is contained in the information. Now we are in search of the knowledge that the information can yield for action, for growth, and for application.

Data to information to knowledge is a perpetual chain in the human search for the meaning of life. Knowledge can be broadly defined as the love of truth. As philosophers search for truth high and low, it is said that they are in a dark room looking for a black cat that isn't there. Theologians also search for truth, and their search is similar to looking in a dark room for a black cat that isn't there, but *they claim* to have found it.

WHAT KNOWLEDGE IS AND HOW IT IS CREATED

Knowledge is love of truth, and everyone pursues it pretty much every day. When we arrive at what we have pursued, we are not completely satisfied, and further pursuit begins. Note that knowledge binds us as much as ignorance because of this nature of pursuit. However, it is necessary to point out that there are lesser truths and greater truths, and we progress from lesser truths to greater truths, and this understanding is needed in pursuing the right knowledge that the organization needs or can use.

According to Ikujiro Nonaka in *The Knowledge Creating Company* (1995), knowledge can be categorized as *tacit* or *explicit*. Tacit knowledge is the highly personal, internalized knowledge of individuals that gives rise to the phrase, "I know more than I am able to tell you," not because one doesn't want to tell, but because one doesn't know how to communicate that knowledge. Explicit knowledge is formal and systematic. It may be codified, easily communicated, and shared.

Based on these two definitions, there are four ways of creating knowledge in an organization:

1. *From tacit to tacit.* An individual directly shares tacit knowledge with another, imparting the tacit knowledge without codifying it. This is commonly seen in the apprentice who observes the master and learns mainly by practice and osmosis. Skill is the body and "feel" is the soul of this knowledge.

2. *From explicit to explicit.* Here an individual collects discreet pieces of explicit knowledge and combines them to create explicit knowledge. Financial statements, research papers, and journal articles are some examples, though not pure interpretations.

3. *From tacit to explicit.* Here an individual with tacit knowledge converts it to explicit knowledge, enabling the sharing of this knowledge with others in a team. This example can also include explicit to explicit knowledge when an individual collects explicit knowledge and adds their own tacit knowledge in creating the codified explicit knowledge.

4. *From explicit to tacit.* When an organization or individual takes in the explicit knowledge and internalizes it over time by broadening, extending, and embedding their own tacit

knowledge, explicit knowledge becomes tacit knowledge. Here is born the familiar organizational culture that says "That's how we do it around here."

In knowledge-creating organizations, all these forms of knowledge exist simultaneously in a spiral of knowledge, that is spiraling out as the organization learns and grows.

KNOWLEDGE-CREATING ORGANIZATIONS

When teachers want to impress on the student the importance of a subject, they tell a story.

A writer was distracted by his active young son, and to keep him occupied, the father took a world map, tore it into many pieces, and gave it to the youngster with instructions to put it back together. (This was obviously before the days of picture puzzles.) Thinking that this task would keep the child busy, the father went about his work. However, in a short time the child was back with the map, fully and correctly assembled. Flabbergasted by his prodigy, the father asked him how he was able to complete the complicated task so quickly. From the mouth of the babe came the reply: "There was a picture of a man on the back of the map, and when I put the man right, the world was alright."

This story demonstrates the importance of knowledge creation and the importance of measurement to manage it. The child's innovation (note that his innocence was a factor—that is, he had no preconceived ideas of the direction to bog him down) enabled his efficiency. If you were to model a learning organization, you would start to assemble the puzzle pieces as it were, and search for the right metrics to track. If the chosen metric were akin to the anatomical parts on the back of the assembled puzzle in our story, then you would be on the right track. Eventually, when you analyzed the metrics, a (human) picture would emerge that would lead to innovation.

The story also contains the truth that when you put the man together right, the world becomes right—that is, when one becomes the change one wants to see, then the organization succeeds ("Be the change you want to see in the world," as

Mohandas K. Gandhi said). The role of leadership in walking the talk is crucial if the organization is to be successful.

LEARNING ORGANIZATIONS

Understanding the serendipity involved with innovation, companies impose a concrete sense of direction by providing fuzzy goals, slogans (which is different from managing by slogans!), and metaphors around which new knowledge can be created. These are identifiable umbrella concepts specific to the organization.

The easy way or the hard way—the learning organization needs to follow certain paths to get to the desirable results. In *Building Learning Organization* (1993) David Garvin of the Harvard Business School identifies three basic issues—meaning or definition, management, and measurement in a learning organization.

As long as an organization has come into being based on some certain knowledge, how does it survive and grow in a fast-changing environment? It does so by organizing itself to create knowledge, and modifying its behavior to use the new knowledge for its own betterment. This definition is organic and identifies the knowledge lifeblood, its function, and its movement in an organization, providing for self-renewal. Some examples of learning organizations include Honda, Canon, GE, NEC, and Kao, where knowledge creation, distribution, and adaptive behavior happen by design, rather than by accident.

KNOWLEDGE CREATION

Learning organizations tap the tacit and very subjective knowledge of individuals and make it available to the whole organization for testing, learning, and adoption. Though the successes of Honda, Sharp, and NEC are many, it should be pointed out that they come from the commitment of individual employees to the business enterprise and its goals. Businesses need to modify and adapt these designs to best fit their environment to make them effective. The cross-functional and self-directed teams are two examples of business attempts to provide an environment to create new knowledge and benefit from knowledge management. A fully committed knowledge management organizational structure is still evolving, though it has been successfully implemented at some companies, including GE and BP.

Some successful companies that utilize knowledge management principles provide the umbrella concepts to the organization for direction. They do not let the competition set the corporate agenda, but look at their customer base and question their existing assumptions to meet their customers' needs.

OBSTACLES TO LEARNING

The leader lives in the future, with the vision for the enterprise; the manager lives in the past, for the certainty it offers toward control; and the technician (the rest of the workforce) lives in the here and now, as today's tasks demand it. Each one looks at the other two with a certain reservation, and this is the first hurdle to learning. The second comes from the refusal to learn from failures, as failure connotes weakness. Single-loop learners who have always been successful and have never had to learn from failure have missed the greatest strength that failure offers. Culturally, we are conditioned to hide failures in general, and a portrayal of strength is paramount. This model may have been born of a pioneer environment. One who does not learn from mistakes tends to be defensive and readily blame others for failures, and so misses the chance to learn.

If this culture is changed to one that values learning from mistakes, one can dramatically improve the learning of an organization. If the 20,000 dispersed technicians know of an error and the circumstances that created it, the organization is that much better off. If management can avoid learning expensive lessons over and over, is that not better for the organization in the long run? Keep in mind that learning from mistakes is *not* the preferred method of learning.

SIX SIGMA AND KNOWLEDGE CREATION

Here is a model of the organizational activities and costs involved in pursuing Six Sigma to make the case for building a knowledge-creating company that answers the innovator's dilemma with knowledge management principles.

Knowledge management is a growing discipline, just as statistical quality control was nearly 60 years ago with pioneers like Walter Shewhart and W. Edwards Deming. Knowledge management had its origins in the last decades of the twentieth century. Though this is not a true beginning in human learning, the demand on the knowledge creation and management process increased with the rate of increase in knowledge creation.

Improvements, Inertia, and Ideality.   Figure 8.10 shows the principles of knowledge management. Applying the proven Six Sigma methodologies, an organization can improve its activities dramatically, achieving dramatic results in measurable ways. Once this level of success is reached, there is the inherent inertia that drags the process changes backward and at best aims for a status quo (we've arrived). The increasing demands of ideality push for further and continuous improvements in Six Sigma efforts as the environment (market, technology, etc.) continuously changes. With the lessons learned from doing Six Sigma projects, the organization must be able to create new knowledge

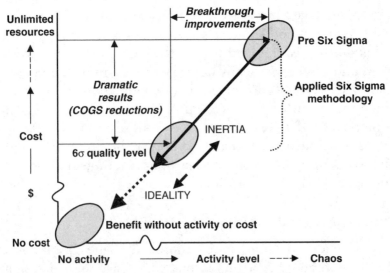

FIGURE 8.10.   Principles of knowledge management.

at a minimal cost. The challenge is to generate more innovation from the innovative solutions that have been developed.

Activities and Costs.   Where is this force of ideality coming from, and how do you nurture and drive it to benefit the knowledge-creating organization? When studying material handling in college, the professor challenged a group of us to define the best way to handle material. As the answers kept building, they all addressed the negative aspects of material handling—and necessarily so, because of the nature of the activity. There was hardly anything positive mentioned about material handling, and *reducing it to a minimum* seemed to be the ideal. So one of our answers was a simple *don't.* That is, *don't* handle the material if it can be helped, because that is the boundary of this activity. Is this possible? Possible or not, it is the ideal that can effectively guide the design of material handling processes. The common saying "We are sitting on a gold mine" can illustrate this point. If we are sitting on a gold mine, all we have to do is securely fence it in and set up a tollgate. Customers can come in, load their buckets, weigh them at the gate, and pay the toll keeper. No material handling!

This is the force of ideality that drives all organizational processes toward *tranquility.* What happens at the boundary is this: all activities cease or slow to a minimum level, and almost all the benefits of the activity are delivered at no cost. It should be cautioned that the utter failure of an organization would also lead to a *sort* of tranquility, with no activity and no cost, but the difference is that there are no benefits, either. This revolution has been demonstrated in the communications industry—the cost of transferring a bit from here to there is near zero, and a few hundred TV channels represent the benefits. (Now, of course, the challenge is the *content.*)

Is this the concept of the lights-out factory? Certainly not. It is the competitive forces of the marketplace that impel companies to force the costs out of business processes but deliver the functionality. A benefit without machines is the ideal concept. Striving toward this ideal creates the benefits in a knowledge-creating knowledge management company.

Why Companies Fail. What can drive this process after applied Six Sigma methodologies have demonstrated the associated benefits? Clayton Christensen outlines the answer at length in his book, *The Innovator's Dilemma* (2003). He discusses how leading companies have failed to sustain their successes, like the farmer who collected the eggs but failed to renew the chicken flocks, to the farmer's eventual downfall. Here the chickens are the process capabilities that need to adapt to new and changing technologies, which provide new benefits to emerging markets and new customers. These customers and markets are necessarily small to begin with, most of the time.

Knowledge-creating organizations go a step further in creating and managing the disruptive technologies that seemingly undermine their own established products and/or services and gradually transition to their leadership positions *continuously*. Reexamining the basic assumptions of the current and successful processes and challenging them successfully, organizations create new knowledge. However, their application may have limited use in the marketplace, and the challenge lies in winning the support of leadership focused on short-term results. Knowledge-creating companies not only encourage such disruptive efforts but also organize the business structure to nurture and harvest them. Lateral-thinking principles can apply here when an organization moves away from the competition and creates *surpetition* (a word coined by Ed DeBono in his book of the same title), whereby the organization is running the race in a league of one. But never for long, as others play catch-up, unless an organization can continuously keep up the superlative behavior mentioned.

Innovation Challenges to Basic Assumptions. Once a trainer was trying to impress us with a new technology being developed that would allow a stranded motorist to walk to the nearest phone and automatically be connected to the nearest *open* repair garage to get help at any time.

"What does this prove?" he triumphantly asked us.

One of us replied, not so softly, "That we need better cars!"

We have come a long way since then. The cars have certainly gotten better, and with cell phones there is no need for roadside phones every few yards. Questioning the assumptions made at all levels of organizational activity will reveal many opportunities to create new knowledge.

Information Responsbility. We already have information-based organizations, except that such organizations are treated as information-processing machines. They have to evolve into knowledge-creating organizations. For everyone to take responsibility for information, we need an organizational structure that encourages information-based management units. As one eagerly looks for the knowledge input, the responsibility for the knowledge output becomes equally important. Based on this, some companies, such as Kao, have initiated "information democracy"—everyone in the organization is granted access to all company information (except personal records). Employees are judged by the contributions they make to the rest of the organization, irrespective of the position they occupy.

This brings us to a serious challenge regarding the enterprise's treatment of its employees as knowledge management takes hold. How do you encourage, nurture, compensate, and value employees based on their knowledge contribution to the company? If two heads are better than one, the combined intellectual capacity of a company's employees is certainly a powerhouse. Since the enterprise is incurring enormous costs in terms of human resources, it is natural to make this aspect of the business activity as productive as others, and harvest the knowledge for the betterment of all in the organization. To begin to assess an organization's knowledge management ability, we need to find out where the expensive lessons learned reside so that no one else repeats the same mistakes.

LEARNING CURVES

As technology improves, we tend to learn more and more about less and less, with the power to create knowledge concentrating at the lower levels of an organization or in the hands of specialists who cannot be told how to do their work. This section attempts to show how a learning organization can create, test, and disseminate such knowledge and know-how for the betterment of all employees. This effort focuses on learning curves and how a study done more than half a century ago can be utilized both in knowledge creation and its utilization.

It was observed that the number of labor hours needed to assemble aircraft decreased on a regular basis as the number of units increased. The second unit required 80 percent as much as the first, and the fourth required 80 percent as much as the second. The concept that emerged was that every doubling in production volume resulted in a 20 percent decrease in labor cost.

The plot of the theoretical relationship between production and labor content is shown in Fig. 8.11. Every doubling of production reduced the labor usage by 20 percent because *learning* was taking place. Knowledge was being created, distributed among the members, and utilized in the operations.

An organization's widely dispersed service teams were studied, applying the insights gained. The learning curves are shown in Fig. 8.12. Team 1, consisting of four members, deploys products or services at customer sites. The team's expended labor time is shown in the shaded area. Learning may take place, but that knowledge would remain within the team if no learning organization and knowledge management structures were in place.

If the deployment involves 1,000 replications of the product or service application, there is an obvious problem (Fig. 8.13). Team 2, deploying units 26 through 50, goes through learning on its own, gains the benefits of this knowledge, and keeps this knowledge, perhaps in tacit form. The benefits of knowledge are not made available by the organizational structure.

When a learning organization is in place and the principles of knowledge management are applied, the learning curve is as shown in Fig. 8.14. Here, when the learning curves of three small teams are plotted together after the principles of a learning organization are applied, we see the benefits of knowledge management. Assuming that the teams work in chronological order, the learning and knowledge transfers to each subsequent team for the benefit of the total deployment effort and the potential savings are considerable.

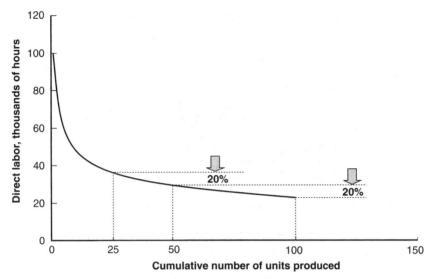

FIGURE 8.11. Learning curve for aircraft assembly.

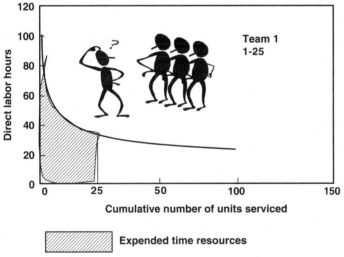

FIGURE 8.12. Time resources expended in learning.

When British Petroleum applied such principles to its oil-drilling operations in far-flung global locations, this principle was proven beyond any doubts. Though this demonstrates the benefit to a learning organization that expends labor, the concepts and principles equally and readily apply to all business

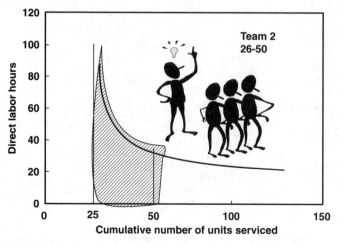

FIGURE 8.13. Learning curve when the knowledge stays with the team.

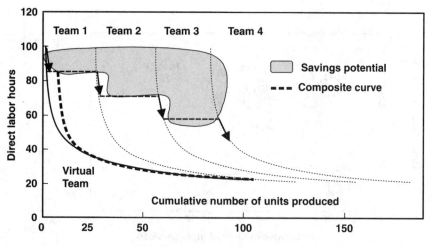

FIGURE 8.14. Learning curve with knowlege sharing.

processes, including professional services and even research and development. The development behind the TRIZ approach to innovation proves how innovation can be brought to practice by applying the scientific principles learned in studying past innovations.

Imagine when each employee is fully informed of other employees' contributions in developing new solutions and creating new knowledge—innovation could become routine. The information management function may become the innovation management function. Innovation is the main value that can be gained from managing information.

Six Sigma has become a leading methodology in the development of breakthrough solutions. During the life of a project, much knowledge is gained from the team's cross-functional experience. Organizations must create a depository to gather all the knowledge gained from each project and processes. This new information could lead to innovative solutions that one cannot even imagine. Creating value from the information generated through experience appears to be the next big thing in our journey of continuous improvement and keeping the competitive juices flowing. The intellectual challenge will be in knowledge management. Companies that implement the knowledge management process will improve at a greater rate, and maintain their growth and profitability. That's the power of knowledge management.

OPTIMIZING SIX SIGMA

Six Sigma is perceived to be an expensive initiative, with a huge potential for return on investment (ROI). However, there are risks associated with it. False starts, lack of commitment, and lack of planning may lead to unsatisfactory results. Implementing Six Sigma successfully is a complex process. Therefore, to minimize the risk of failure and improve ROI, one should treat Six Sigma as a process that needs to be improved for better, faster, and cheaper performance. In other words, the process of implementing Six Sigma must be improved and optimized.

Since no one has applied or improved Six Sigma from that angle, this is an opportunity to explore the effect of the Theory of Constraints (TOC) on a Six Sigma process that is implemented at the corporate level. TOC is an industry-recognized methodology for improvement.

OVERVIEW OF THE THEORY OF CONSTRAINTS (TOC)

Eli Goldratt developed an information system called Optimized Production Technology (OPT) in the late 1970s through the early 1980s. The management philosophy of TOC was developed in the late 1980s, primarily from his work. TOC examines the thinking processes of individuals who invent simple solutions and the psychology of an organization in adapting to those solutions. The OPT tools for thinking evolved into the more generic and far-reaching concept of TOC. The following basic principles constitute TOC:

1. Organizations have goals.
2. Any organization is more than the sum of its parts.
3. A few vital variables constrain an organization's performance.

The system thinking expressed by the second principle leads to the need to synchronize resources and understand the interdependencies of various parts of any organization for a common purpose, invariably to meet the needs of customers.

Anything that limits the performance of an organization can be seen as a constraint, and only a few of them are active at any given time, limiting the development of an inventive solution. There are five steps, or areas of focus, in identifying and overcoming the constraints and realizing improvement (Fig. 9.1).

1. *Identify the constraints.*   Identifying the system constraints and prioritizing them so as to progress toward the goal is the first step. In other words, what is it you do not have enough of to reach your goals, or which constraints are preventing you from finding the right solution? Putting your finger on this limited resource that restricts the entire system is critical if your efforts to improve the process is to be effective.

2. *Exploit the constraints.*   The next step is exploring and deciding how to exploit the constraint. This includes managing the constraint, by providing just what is needed to maximize the output of the constrained resource. Do not let the output of the upstream resources exceed what is needed by the constrained resource. Doing so does not move the organization closer to its goal.

3. *Subordinate the constraints.*   Subordinating all other things to the preceding decisions is the third step. This prevents the supplying processes from accumulating their output before it is needed by the constrained one. Since constraints exist for a

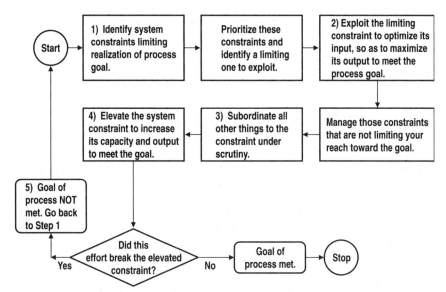

FIGURE 9.1.   Theory of constraints process map.

reason, it is possible to do something to modify them, reducing their limiting effects.

4. *Elevate the system's constraints.*   Elevating the system's constraints is next. That is, if you have exploited the constrained process, and the demand on its output is still not met, you need to find other ways to increase its capacity. Elevating the constraint can eventually break it, and another constraint may emerge as the limiting factor in the process.

5. *Repeat Steps 1 to 4.*   If the first constraint you started with is broken, start at Step 1 again, without allowing inertia to cause a constraint. Inertia comes from assumptions not examined, existing policy restrictions whose justification has long since vanished.

The TOC is not a special quantitative method but a logical algorithm that lends itself to incremental application in any process. It may lead to suboptimization initially, but continuous and sustained effort eventually moves the organization toward its goals, consistently delivering improvements. Ultimately, the objective in applying TOC is to understand the problem, the solution, and the implementation of the solution using the employees' knowledge and the organization's collective psychology. Figure 9.2 summarizes the steps of the TOC method.

TOC AND PDCA MAPPING

W. Edwards Deming defined the Plan, Do, Check, and Act (PDCA) model for achieving effective results in any activity or process. An analysis of the TOC method with respect to Deming's PDCA model is shown in Fig. 9.3. The first focus area of TOC, identifying system constraints, is akin to the planning aspect in optimizing constraints. Just as in setting up a process or planning for the right input, identifying the constraints establishes the scope of the process for which the constraints need to be optimized. Identifying constraints is a process similar to cause-and-effect analysis, where variables

TOC STEP	DESCRIPTION
1. Identify system constraints.	Identify those limited resources you do not have enough of to reach your goal.
2. Exploit those constraints.	Explore how to manage the constraint by providing the input needed to maximize its output.
3. Subordinate all other things to the above decision.	Manage those constraints that are limiting. This requires demotion of the processes of factors that constrain the desired goal.
4. Elevate the system's constraint.	If after exploiting the constrained process, its output does not meet the demand, find other ways to increase its capacity and output. If this constraint is broken, another constraint may appear as a limiting factor.
5. If the selected constraint is broken, go back and start at Step 1.	A constraint is broken when it is no longer a hurdle toward the goal, but the goal is not met. Now start at Step 1 again, and identify another constraint.

FIGURE 9.2.   Summary of TOC steps.

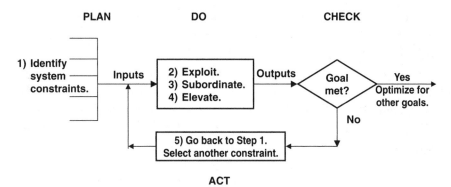

FIGURE 9.3.   TOC and PDCA mapping.

are identified in the areas of materials, machines, methods, and people. Similarly, according to TOC, you identify all the system constraints that are involved in enabling or restricting the goal of the process or organization.

The Do phase of the PDCA model consists of a set of activities that one pursues to produce the desired results. Three of the steps in TOC fall under the Do phase:

• Exploit the constraints.
• Subordinate the constraints.
• Elevate the system's constraints.

After executing the operational steps, verification of the output against the objectives is performed. This part of any process is the Check phase, which includes many details concerning metrics, what is acceptable, what is not acceptable, and degree of conformance, or achievement against the goal. According to the TOC model, in this phase you check to see whether the exploitation and elevation of the identified constraint have enabled the process to meet its intended goal of production or service. If the answer is yes, you have achieved the goal set for the process. If the answer is no, the constraint has been broken, but you have not achieved the process goal. Interestingly, this decision-making step is not included in the five focus areas of TOC.

The Act phase of PDCA is actually a correction process. The PDCA model requires remedying the deficiencies of any process. The TOC focus area of repeating the steps is slightly different from adjusting a process deviation. According to TOC, this step enables the identification of another constraint and logically pulls it through the TOC objectives. According to PDCA, the Act step remedies the deficiency, while the TOC launches the process of continuous improvement.

EXPLOITING TOC CONCEPTS FOR SIX SIGMA

TOC is a process for increasing an organization's output toward its goal by institutionalizing an optimized solution. Goldratt identified five thinking-process tools used in this theoretical framework to identify need, the opportunity to change, and potential solutions to affect the bottom line positively. Descriptions of specific tools mentioned further into this treatment follow.

Current Reality Tree (CRT).   This tool describes the capture of the experience and perceptions of the owners or stakeholders of the process. It may even identify the root causes of the core problems of the organization. The core sufficiency-based logic is "if…, then…." This helps the team to understand why X leads to Y and to insert new actions, including replacing X or changing the result to be Z as required. The CRT is also a communication tool, prepared by subject matter experts and used for examination and clarification in presentations to the team. Team members can also use this effectively to build their part of the process, so that the full team can assemble the total picture of the current process. CRT and the evaporating cloud tool link the range of issues faced by the team and provide the answer to the question of what to change. The CRT equates to establishing a baseline in terms of perceptions and symptoms.

Evaporating Cloud.   This tool is a key contribution to the thinking process, much utilized in TOC as well as in negotiation and conflict resolution processes. This thinking tool helps to identify contradictions that are at the source of a problem, and hence is also known as a *conflict cloud resolution diagram.* The main purpose of this tool is to identify the source of conflict or impasse that can lead to a win-win solution to a core problem, since no one in the organization is in conflict with the goal and purpose of the organization (and even this can come under examination when necessary). The process involves the examination of underlying assumptions, which may have been justified long ago, when they were adopted. Examination and elimination of these assumptions evaporates the conflict. Though this works well as a stand-alone tool, it is also integrated into the CRT. The evaporating cloud could be used to identify conflicts instead of causes, assuming that there is more than one variable causing the problem.

Figure 9.4 shows that to achieve objective A, two requirements, B and C, must be met. To meet requirement B, prerequisite D must be met, and to meet requirement C, prerequisite D' must be met. Prerequisites D and D' have conflicts and must be resolved. An optimization between D and D' must be achieved. Every arrow in the figure is based on one more underlying assumption that one needs to scrutinize consistently.

Figure 9.5 shows the application of TOC elements to cost reduction in the manufacturing environment. Reaching objective A (reducing cost) requires reductions in setup costs and inventory carrying costs. The setup cost per part can be reduced by enlarging the batch size, while the cost of carrying

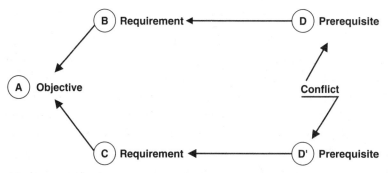

FIGURE 9.4.   The evaporating cloud.

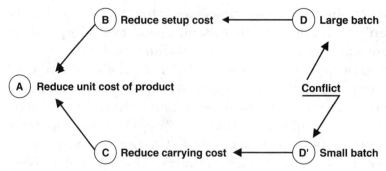

FIGURE 9.5. Applying TOC to cost reduction.

inventory can be reduced by reducing the batch size. It appears that the two constraints in reducing unit cost lead to conflicting requirements that must be resolved.

Future Reality Tree (FRT).   This tool helps to identify what is missing from a solution. This is a necessary step of evaluating and improving a solution prior to implementation. The FRT modifies the CRT with new actions inserted into it, so as to create the future reality for the system. The process of "if…, then…" removes a cause and shows the results as they are affected. An iterative application of such logic helps the team create new solutions in this evaluation step. The final objective of the FRT is to communicate a vision of desirable effects when undesirable ones identified with the CRT are changed. This step essentially helps the team to answer the question of what to change to. Figure 9.6 shows an example of the CRT transforming into three alternative FRTs in developing an approach to fighting a forest fire.

Prerequisite Tree (PRT).   This thinking tool identifies the transitional steps needed to attain the solution. This step is more like answering the question of why this cannot be accomplished. The answers form an exhaustive list of hurdles preventing the chosen results from being achieved. This is an energetic step for the team, because the natural and hidden fears of those involved will be brought out to be addressed. Otherwise, these fears would prevent or retard the final deployment. The PRT also creates the action responses that will help overcome the negatives,

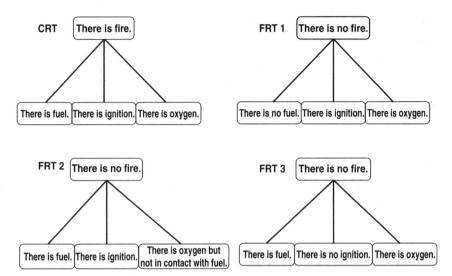

FIGURE 9.6.   Current and alternate future reality trees (FRTs).

and the priorities for addressing them. The results or degree of success can be predicted based on the success of these prioritized actions. The FMEA could be used to construct a PRT.

Transition Reality Tree (TRT).   Once the transitional steps are identified with the prerequisite tree, this tool helps to identify specific actions needed in the current situation. This is more or less a cause-and-effect relationship summary describing the team's course of action identified earlier. This cause-and-effect relationship is used for forward analysis instead of feedback analysis. It provides a clear picture of how the actions to be taken combine with the existing reality, yielding the desired outcome planned.

All the tools described can be used as stand-alone tools or in integrated combinations for problem solving and conflict resolution. As described, the team is fully engaged in identifying conflicts and in resolving them. This ensures their buy-in of the solution and helps the team effectively solve a number of problems by bringing in needed information, perhaps hidden away in various parts of the organization.

The following list summarizes the five thinking tools:

TOC Thinking Tools	Conventional Tools
Current reality tree	Baseline
Evaporating cloud	Understanding and resolving contradictions
Future reality tree	Establishing goals
Prerequisite tree	FMEA/validation to prevent obstacles
Transitional tree	Action plan to achieve goals

It appears that the evaporating cloud and the prerequisite tree are two excellent steps for resolving conflicting constraints and preventing potential obstacles from occurring in order to achieve desired goals.

IMPLEMENTATION

The implementation of TOC requires understanding what to change, what to change to, and how to make the change happen, and TOC provides appropriate logical tools to deal with these questions. The TOC method is more qualitative than quantitative. It is more a mental process than an analytical tool. The tools focus on relationships more than quantitative aspects in analyzing a problem.

What to Change.   A key part of any change initiative is knowing what to change. This is covered in Step 1 of the TOC process. The analysis of the current status is formed by the assessment of the situation, a description of the current reality, the identification of the core problem, and the assumptions that are sustaining the problem. This process may include the voice of the customer, employees, and suppliers, for their understanding, their perceptions, and their feedback.

  Applicable TOC tools.   The current reality tree and evaporating cloud steps are used to identify the areas of focus for improvement.

What to Change To. The main part of the TOC effort involves describing the vision and strategy to attain the goal, or the solution to be developed.

  Applicable TOC tools.   The evaporating cloud for an innovative solution, and the future reality tree to show the desired state.

How to Make the Change Happen. This phase covers the detailed plans and tactics for managing the implementation of the changes through proper planning and teamwork.

Applicable TOC tools. The prerequisite tree and the transition tree are used to validate the solution and develop an action plan to achieve the future tree state.

MANAGING LAYERS OF RESISTANCE TO CHANGE

In the implementation phase of TOC, Goldratt recognized various layers of resistance to change that can be expected to occur in any organization. These layers of resistance are commonly experienced while achieving the future tree state. The layers of resistance to change are as follows:

Disagreement about the problem. Conflicting priorities and perceptions lead to this resistance. Employees have many things to do. The problem to be addressed must be based on common experience or be systemwide.

Disagreement about the direction for a solution. Employees have ideas about solving the problem based on their experience and expertise. The approach to solving the problem must be based on employee input instead of just the team leader's or manager's input.

Disagreement about the solution. Given that TOC is a thinking approach instead of a quantitative method, gaining the consensus of employees involved in solving the problem will minimize the disagreement about the solution. People disagree because the solution does not fit their experience model. Therefore, for employees to accept a solution, it must agree with their logic of thinking, or a different understanding must be developed through communication or training.

Uncertainties about adverse side effects of the solution. Employees do care about the company and the solutions they implement. However, they are cautious of adverse affects the change may bring to the company. The solution must be discussed, and the prerequisite tree could be used to anticipate

404   CHAPTER NINE

and resolve potential problems with the solutions.

Barriers to change for personal reasons.   Over time, people develop certain behaviors and practices that they believe are personal. They must be clear that the solution is not about changing them or their behaviors; instead, the change is about the process they follow.

Fears not verbalized.   Some people are just quiet, or sometimes people are afraid due to heavy-handed management practices. People are afraid of retribution or reprisals by management if they share their personal concerns. An open and consistent management process will help to alleviate employee fear. Recognition of employees for their input will open up employees even more regarding their ideas and recommendations, and make them more friendly to change.

APPLYING TOC TO SIX SIGMA

Six Sigma evolved from Motorola's need to maintain its leadership position and accelerate its continual success by proactively addressing customer concerns. In the mid-1980s, quality was a number-one customer concern. It still is; however, the bar has been raised. During the past 15 years, all industries have addressed quality issues, and this has become an assumed customer requirement. The next phase of improvement leads to dramatic improvement, instead of continual improvement. Motorola used Six Sigma to renew and reenergize the organization by setting big goals. Since then, Six Sigma has evolved into a well-defined system of performance improvement. The Six Sigma methodology and tools have become standard objects. The Six Sigma methodology has been successful to a large extent, but some inefficiencies and bureaucracy have crept in. Such conditions have led some companies to abandon their Six Sigma initiatives and discouraged other companies from attempting to benefit from the power of Six Sigma. The key component that makes Six Sigma successful is the organizational psychology. In other words, the success of Six Sigma depends to a great extent on Six Sigma thinking, or on the soft skills of the organization besides the fact-oriented statistical tools. These factors are hard to standardize and quantify. Therefore, the application of TOC as shown in Fig. 9.7 appears to be a great tool for improving the performance of the Six Sigma system for better results.

In a typical Six Sigma implementation, the executive team makes the decision about implementation based on its potential for solving some set of problems. If the problems are solved, the corporate leadership makes the decision to institutionalize Six Sigma. A Master Black Belt or a Black Belt is hired to lead the Six Sigma initiative. The training resources are unleashed to provide training to employees in a specified time. Then, some trained people are assigned to projects and others wait for their turns. Management implements an extensive project management program to track projects. Some companies keep the Six Sigma internally focused, and some focus on customers. In other words, some decide not to publicize it, while others publicize it for marketing purposes. In some companies the CEO gets in the driver's seat, while in others the Six Sigma driver is hired. With all these variations, a few success stories are publicized, and many struggles are ignored, along with the lessons that could be learned from the failures. Sometimes the distinction between failure and success is blurry, either intentionally or inadvertently.

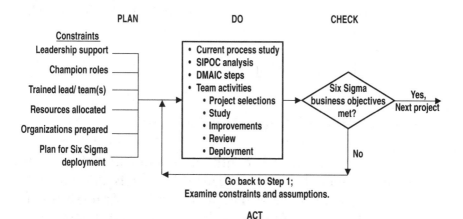

FIGURE 9.7. Application of TOC and PDCA to Six Sigma.

In applying TOC to the Six Sigma implementation process, the first step is to identify constraints. Constraints are variables that can affect the success of the Six Sigma initiative. The constraints can be listed based on the process flow and the four basic elements of a process: material or information, method or approach, equipment or tools, and people or skills (Fig. 9.8).

EXPLOITING AND SUBORDINATING CONSTRAINTS FOR SIX SIGMA

After identifying the constraints to implementation of the Six Sigma initiative, one must maximize the positives and minimize negative constraints. Figures 9.9 to 9.12 list constraints and assess their effects in either *supportive* or *adverse* categories. A constraint can be supportive or adverse according to the following guidelines:

Item	*Supportive or Adverse*
Material or information	Critical or noncritical
Approach	Supportive or adverse
Equipment or tools	Needed or not needed
People (resources)	Improve or reduce

In order to break the constraints, one must look at them from different perspectives and for higher objectives. The higher the level one looks at constraints from, the more a theme starts to evolve, as multiple constraints relate to a common higher-level constraint.

DEVELOPING AN ACTION PLAN

Once the constraints are identified, exploited, and subordinated, an action plan or the transition tree is developed to achieve the goal of successful implementation of the Six Sigma initiative. The action plan for exploited and subordinated constraints is shown in Figs. 9.13 and 9.14.

PEOPLE	EQUIPMENT OR TOOLS	MATERIAL OR APPROACH	INFORMATION
Qualified resources	Executive team training	Strategy for implementing Six Sigma	Leadership education
Internally available, enthusiastic individual with strong leadership and moderate statistical skills	Time management skills training	Corporate performance measurement model (e.g., Six Sigma Business Scorecard or Balanced Scorecard)	Drivers for Six Sigma
Experienced Black Belt for mentoring	Innovative thinking training	Performance communication with stakeholders	Market position
Qualified candidates for Black Belt expertise (buy or build?)	Six Sigma awareness training	Fair and objective performance review system	Business opportunity analysis
Project mentors from successful project teams	Statistical thinking training	Commitment to communicate consequences of poor performance	Competitive assessment and awareness
Expanded Six Sigma capability based on needs	Green Belt training	Commitment to reward and recognize excellence	Corporate performance scorecard
Project teams	Six Sigma software	Continual assessment and renewal	Business opportunity analysis
	Information management systems	Recognition process	Prioritized list of projects
		Communication of corporate vision, values, and goals	Organization (ownership) chart

FIGURE 9.8. Constraints for Six Sigma.

PEOPLE	EQUIPMENT OR TOOLS	MATERIAL OR APPROACH	INFORMATION
			Measurements of success
			Departmental measurements
			Competitive compensation information
			Financial results
			Successful completion of projects with significant savings
			Innovative solutions

FIGURE 9.8.   Constraints for Six Sigma. (*Continued*)

FOLLOWING UP ON THE ACTION PLAN

After the action plan has been implemented, one must review progress toward the established goals and make necessary adjustments. The Six Sigma initiative again must be reviewed for bottlenecks or barriers to progress, be they the in-house expertise, cost of training, or availability of qualified resources. The implementation plan must reflect changes to reflect achievement of desired goals.

THE OPTIMIZED SIX SIGMA IMPLEMENTATION PLAN

The supplier, input, process, output, and customer (SIPOC) method is a great tool for identifying various players and related information in the Six Sigma theatre of operation.

Constraint	Effect (Supportive or Adverse)	Elevation	Subordination
Leadership education	Supportive	Leadership's understanding	
Drivers for Six Sigma	Supportive	Leadership's understanding	
Business opportunity analysis	Supportive	Leadership's understanding	
Corporate performance scorecard	Supportive	Scorecard	
Prioritized list of projects	Supportive	Leadership's understanding	
Organization (ownership) chart	Supportive	Leadership's understanding	
Measurements of success	Supportive	Scorecard	
Successful completion of projects with significant savings	Supportive	Scorecard	
Innovative solutions	Supportive	Beliefs and culture	
Market position	Adverse		Internal improvement focus
Competitive assessment and awareness	Adverse		Internal improvement focus
Departmental measurements	Adverse		Internal improvement focus
Competitive compensation information	Adverse		Recognition
Financial results	Adverse		Scorecard

FIGURE 9.9.   Material or information constraints.

CONSTRAINT	EFFECT (SUPPORTIVE OR ADVERSE)	ELEVATION	SUBORDINATION
Strategy for implementing Six Sigma	Supportive	Planning	
Corporate performance measurement model (e.g., Six Sigma Business Scorecard or Balanced Scorecard)	Supportive	Scorecard	
Performance communication with stakeholders	Supportive	Communication	
Fair and objective performance review system	Supportive	Performance System	
Commitment to communicate consequences of poor performance	Supportive	Communication	
Commitment to reward and recognize excellence	Supportive	Recognition	
Continual assessment and renewal	Supportive	Scorecard	
Recognition process	Supportive	Recognition	
Communication of corporate vision, values, and goals	Supportive	Communication	

FIGURE 9.10. Approach constraints.

SIPOC utilizes the process steps, inputs to the process steps, and their source. On the other side, SIPOC identifies the process outputs and their destination for a better understanding of customer input, interface, and interests.

SIPOC incorporates the optimized process flow for cost-effective implementation of Six Sigma. The process steps in the SIPOC model reflect modifications to reduce cost and improve

Constraint	Effect (Supportive or Adverse)	Elevation	Subordination
Executive team training	Supportive	Training	
Time management skills training	Supportive	Training	
Innovative thinking training	Supportive	Innovation	
Six Sigma awareness training	Supportive	Training	
Statistical thinking training	Supportive	Training	
Green Belt training	Supportive	DMAIC (benefit)	
Black Belt training	Adverse		Tools (cost)
Six Sigma software	Adverse		Tool (cost)
Information management systems	Adverse	Tool (cost)	

FIGURE 9.11.   Equipment or tools constraints.

Constraint	Effect (Supportive or Adverse)	Elevation	Subordination
Internally available, enthusiastic individual with strong leadership and moderate statistical skills	Supportive	Motivating environment	
Qualified candidates for Black Belt expertise (buy or build?)	Supportive	Motivating environment	
Project mentors from successful project teams	Supportive	Motivating environment	
Qualified resources	Adverse		Resources (cost)
Experienced Black Belt for mentoring	Adverse		Resources (cost/guidance)
Expanded Six Sigma capability based on needs	Adverse		Resources (cost)
Project teams	Adverse		Resources (cost)

FIGURE 9.12.   People constraints.

ELEVATED CONSTRAINT	ACTION PLAN TO ACHIEVE GOALS
Leadership's understanding	Understand drivers, opportunity, scorecard, and values development.
Scorecard	Implement departmental and corporate measurements for accelerating rate of improvement.
Beliefs and culture	Establish corporate beliefs and create value-based culture.
Planning	Establish goals and plan to achieve goals.
Communication	Communicate with stakeholders, employees, customers, and suppliers.
Performance system	Establish employee performance system for guidance and skills development.
Recognition	Inspire people for achieving success through frequent recognition.
Basic training	Provide basic innovation and statistical and Six Sigma training.
Innovation	Establish expectations and process for innovative solutions.
Green Belt training	Provide project-driven Green Belt training as necessary.
Motivating environment	Reward superior performance and encourage positive behaviors.

FIGURE 9.13. Action plan (transition tree) for elevated constraints.

performance (Fig. 9.15). For example, if a CEO commits to Six Sigma, SIPOC shows what input will be needed for the commitment to become real. Similarly, planning for Six Sigma includes inputs such as a prioritized list of projects, executive team training, a strategy for implementing Six Sigma, resources for training and mentoring, measurements of success, and an organization (ownership) chart. Another modification to the Six Sigma approach would be to implement a corporate performance measurement system. An effective scorecard is implemented with inputs such as the corporate performance measurement model (e.g., the Six Sigma Business Scorecard or Balanced Scorecard), departmental measurements, performance review process and frequency, and performance communication with stakeholders. A well- implemented scorecard

SUBORDINATED CONSTRAINTS	ACTION PLAN TO ACHIEVE GOALS
Internal improvement focus	Instead of depending on too much external information to implement Six Sigma, establish internal focus for improvement to launch Six Sigma initiative.
Financial results	Minimize cost of Six Sigma initiative.
Recognition	Compensate employees for extraordinary performance.
Black Belt training	Train qualified people to develop in-house expertise.
Six Sigma software	Buy based on concurrent improvement projects.
Information management system	Report more accurate information and analysis periodically.
Qualified resources	Identify people for Six Sigma leadership based on their interest, aptitude, skills, and utilization.

FIGURE 9.14.   Action plan (transition tree) for subordinated constraints.

is a prerequisite for sustaining the Six Sigma initiative over a longer term. Besides the scorecard, the leadership must renew the Six Sigma initiative with some exciting changes, energy, and activities.

The SIPOC in Fig. 9.15 can be modified to suit a company's requirements and business objectives. However, it provides a good framework for understanding and implementing the Six Sigma initiative as well.

MANAGING CONSTRAINTS FOR OPTIMIZING RESULTS

One can see that in implementing the Six Sigma initiative, there are many constraints that must be managed well to optimize results. Just having everyone become a Black Belt or a Green Belt would not suffice in achieving the desired results. There is more to Six Sigma than the training. The most significant factor is the leadership's understanding of the Six Sigma

SUPPLIER	INPUT	PROCESS	OUTPUT	CUSTOMER
Right training provider	Leadership education	Commit to Six Sigma.	Commitment to corporate growth and profitability through Six Sigma	All stakeholders: Customers Employees Owners
Performance information	Drivers for Six Sigma		Revised vision and direction	Management Shareholders
Market intelligence	Market position		Expected financial gains	
Decision to perform business opportunity analysis	Business opportunity analysis			
Candidates to drive improvement	Qualified resources			
Resources for competitive information	Competitive assessment and awareness			
Corporate performance system	Corporate performance			
Management	Internally available, enthusiastic individual with strong leadership and moderate statistical skills	Appoint a Six Sigma corporate leader.	A performance-driven Six Sigma leader	Corporation and stakeholders
Management	Business opportunity analysis	Identify key areas for profitability improvement	List of projects Cost-benefit analysis Impact on profitability Opportunities for growth	Management Six Sigma champions

Management	Prioritized list of projects	Plan for Six Sigma.	Understanding of corporate goals and approach	Middle managers Employees Customers Suppliers
Training provider	Executive team training Strategy for implementing Six Sigma Resources for training and mentoring Measurements of success Organization (ownership chart)		Change agents Six Sigma implementation plan	
Performance model provider	Corporate performance measurement model (e.g., Six Sigma Business Scorecard of Balanced Scorecard)	Implement performance measurement system for the Six Sigma initiative.	Corporate report card Communication methods and schedule Performance trends Actions to achieve desired results	Management Six Sigma leaders in the company Employees
Management (senior and departmental)	Departmental measurements Performance review process and frequency Performance communication with stakeholders			

FIGURE 9.15. Six Sigma SIPOC.

Supplier	Input	Process	Output	Customer
Management	Competitive compensation information Commitment to reward and recognize excellence Commitment to communicate consequences of poor performance A fair and objective performance review system	Establish a Six Sigma performance-driven compensation system.	Clear communication of corporate expectations and recognition system for employees, including management, Six Sigma leaders, and employes	Employees
Management Statistical and Six Sigma training resource Innovation training resource Time management skills training	Communication of corporate vision, values, and goals Statistical thinking training Six Sigma awareness training Innovative thinking training Time management skills training	Conduct basic training for employees.	Basic skills in interpreting data correctly, Six Sigma methodology, innovation process, and prioritizing and executing tasks	All employees
Management Green Belt training provider Black Belt provider	Projects for improvement Green Belt training Experienced Black Belt for mentoring	Conduct project-driven Six Sigma training.	Trained employees Successfully completed projects	Corporation and stakeholders
Departmental managers	Departmental measurements Corporate scorecard Financial results	Verify quantifiable impact on bottom line because of preceding projects.	Savings and improvement in bottom line	Corporation and stakeholders

Supplier	Input	Process	Output	Customer
Six Sigma project teams Employees Management	Successful completion of projects with significant savings Innovative solutions Extraordinary effort and results Recognition process	Recognize success.	Inspirational recognition of employees for their accomplishment Positive work environment	Employees
Human resource department Management Black Belt training provider	Qualified candidates for Black Belt expertise (buy or build?) Project mentors from successful project teams Lessons learned	Develop internal expertise and resources.	Cadre of Six Sigma professionals suitable for the current organization	Corporation and stakeholders
Six Sigma champions Management	New projects Project teams Expanded Six Sigma capability based on needs	Institutionalize Six Sigma methodology.	More savings Savings sharing with employees	Corporation and stakeholders
Management Employees	Scorecard Continual assessment and renewal	Monitor and sustain improvement.	Continual successes Growth Profitability	Corporation and stakeholders

FIGURE 9.15. Six Sigma SIPOC. (*Continued*)

process. It is critical that the leadership understand the intent, thinking, methodology, measurements, and tools of Six Sigma.

Figure 9.16 lists various constraints that are identified, exploited, and subordinated or balanced using the TOC approach. The factors that need to be exploited include leadership's understanding that Six Sigma can be maximized. Interestingly, most of these factors are approach or process related. The emphasis must be on establishing the right process and executing it to the fullest. On the other hand, for the factors that need to be subordinated or balanced, creativity using the evaporating cloud concept must be the focus. The purpose is not to eliminate these factors; instead discretion must be applied to establish their level. For example, Black Belt training must not be eliminated; instead, it must be provided to the right people (attitude and aptitude) and in right numbers based on the opportunity. The creativity could be applied in using the Black Belt training instead of throwing bodies at it. Another factor is the cost of implementing Six Sigma. Many

FACTORS TO BE EXPLOITED	FACTORS TO BE BALANCED
Leadership's understanding	External drivers (include internal drivers)
Scorecard	Cost (reduce)
Beliefs and culture	Compensation (justified)
Planning	Black Belt training (as needed)
Communication	Software tools (appropriate emphasis)
Performance system	Committed resources (commensurate with cost–benefit analysis)
Recognition	
Basic training	
Innovation	
Green Belt training	
Motivating environment	

FIGURE 9.16. Summary of factors to be exploited or balanced.

companies in the glamour of Six Sigma do not track the costs associated with it. But like any investment in the business, there must be an expectation of return on investment that can justify the cost. Finally, the drivers for the Six Sigma initiative must not be purely the external requirements. The key drivers for implementing Six Sigma must be profitability and growth.

SPEEDING UP SIX SIGMA

It has always been the quest of the competitive world to reach markets before others do and to generate wealth. That means the keen desire to achieve results better and faster. The Lean systems provide a way for an organization to leverage its resources in order to improve customer satisfaction and profitability. In implementing Six Sigma solutions, it is of the utmost importance that the problems solved by using Six Sigma methodologies *remain* solved. Sustaining the gain is equally as important as implementing the solution. The effectiveness of a solution is often measured by how the solution continually creates superior performance. Certain Lean tools provide robustness to solutions that are determined to be effective in solving a problem or improving a process outcome. Lean aligns well with Six Sigma, as both represent process-centered, customer-focused, and fact-based approaches. The broad organizational thinking in the three environments—conventional, Six Sigma, and Lean—are compared in Fig. 10.1.

EARLIER EFFORTS TO REDUCE CYCLE TIME

Efforts to speed up processes were first evidenced in the early 1800s when Eli Whitney promoted the idea of interchangeable parts. As a result, he was able to sell muskets to the U.S. Army at unbelievably low prices. During the early 1900s, Fredrick

ASPECT	CONVENTIONAL	SIX SIGMA	LEAN
Management	Cost and time	Quality and time	Speed and flow
Process adjustment	Tweaking	Statistical controls	Mistake proofed
Problems	Repair	Prevention	Elimination
Analysis	Experience	Data based	Data based
Focus	Product	Process	Customer value
Supplier selection	Cost and time	Process capability	Flow capability
Outlook	Short term	Long term	Longer term
Decision making	Intuition	Fact based	Fact based
Organization	Authority	Learning	Autonomy
Training	Convenience	Necessity	Natural element
Direction	Ad hoc	Planned	Planned
Goal setting	Realistic perception	Reach out and stretch	Producing to demand
Control	Centralized	Distributed	Localized

FIGURE 10.1.   Comparison of conventional Six Sigma, and Lean Organizational thinking.

Taylor invented what was termed *scientific management,* developing work methods and introducing concepts such as time study and standardized work. Frank Gilbreth and Lillian Gilbreth introduced motion study, process charting, value-adding and non-value-adding activities, and the motivation and attitudes of workers. Together they established the premise of *waste elimination,* which forms a key element of modern Lean thinking. In Six Sigma terms, waste is also related to excess variation, affecting both one's confidence in the process and the overall cost of process output.

Throughout the early history of manufacturing, product making was perceived as a discrete process wherein each operation was performed independent of other activities or successive operations. The work was done with little attention to work methods, product movement, or process sequencing.

This all began to change in the early 1900s with one individual—Henry Ford. Ford, along with Charles Sorensen, is credited with establishing the first *manufacturing system*. In establishing their manufacturing system, Ford and Sorensen considered key inputs such as people, machines, materials, products, and procedures. Ford thus was credited as being the first practitioner of the just-in-time and Lean concepts.

However, the Lean concept as we know it today apparently started with the Toyota Production System. After World War II, Japanese companies began to study American manufacturing systems. Taichii Ohno and his associate Shigeo Shingo at Toyota began to integrate Ford production, statistical process control, and other methodologies into their system and named it the Toyota Production System (TPS). To some, it was also known as the *just-in-time* (JIT) system. It was recognized early on in the development of TPS that inventory played a central role in driving human behavior and that people played a key role in making the system work. As the system evolved, other elements and concepts were introduced. Terms such as *product variety, kanban, quality circle, cellular manufacturing, batch sizing,* and *setup reduction* were adopted by the TPS and JIT models.

TPS proved a benchmark model for success in manufacturing. Many companies that tried to emulate the TPS model did not do so very successfully, as they failed to understand the underlying principles and philosophies. By the early 1980s some companies such as Kawasaki and GE were able to emulate the TPS model more successfully. With success sprouted a lot of writing and consulting activity, and many acronyms, such as *world-class manufacturing* (WCM), *continuous flow manufacturing* (CFM), *one-piece flow*, and the like became synonymous with JIT production.

Lean manufacturing, Lean thinking, or Lean enterprise as we know it today became popular in the wake of James Womack's 1991 book, *The Machine That Changed the World*. Womack described in his book the basis of Lean thinking as practiced by some of the best companies in the world. In his second book, *Lean Thinking: Banish Waste and Create Wealth in Your Organization* (1996), Womack described how a mass production plant can become a Lean organization. Today Lean

methodologies are identified as key success factors for major
U.S. and international corporations.

INTEGRATION OF LEAN INTO SIX SIGMA

Lean and Six Sigma have long been discussed independently
and only recently have their strengths been brought together. It
is important to recognize the linkage of Lean and Six Sigma, as
shown in Fig. 10.2. A brief overview of the key features of both
systems is depicted in Fig. 10.3.

Lean tools may be applied in all phases of the Six Sigma
approach. These tools help in sustaining the gains and achievements initiated by Six Sigma. The relevant Lean and Six Sigma
tools are depicted in Fig. 10.4. Many companies, big and small,
local and global, have applied Lean and Six Sigma successfully.
Some have made them a natural process of doing business,
while others use them as tools in improving selected processes.

Lean was first perceived as a tool for improving manufacturing processes. However, with further evolution and understanding of the concepts of processes, inputs and outputs, and
value added, Lean was shown to be applicable to all aspects
and processes of any organization. Thus, it is now equally
applicable to manufacturing and service industries. Service
industries such as transport, warehousing, accounting, and

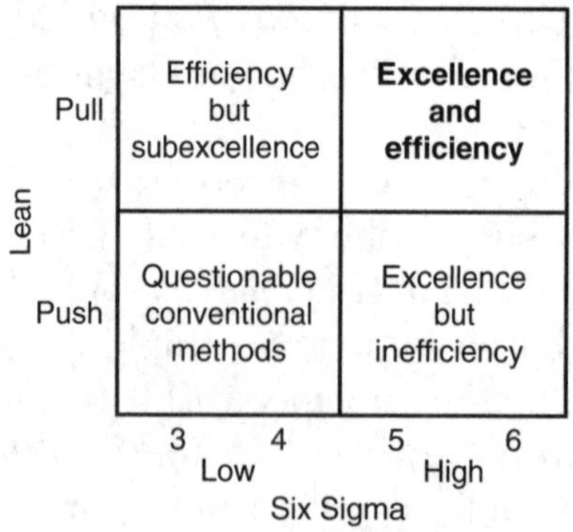

FIGURE 10.2.   Six Sigma and Lean matrix.

Six Sigma	Lean
Typically driven by leadership	Typically driven by middle management
Provides a clear focus and target	Supports the Six Sigma focus and target
Requires dramatic improvement through innovation	Results in continual reduction in cycle time to balance with customer demand
Requires passionate and inspirational commitment from the CEO to achieve perfection	Requires personal commitment to challenge current processes
Affects all aspects of business products and processes	Affects selected processes for speed and value
Driven for excellence	Driven for efficiency
Difficult to implement and benefit from Six Sigma in a localized area	Can be implemented locally in any operation
Many tools and DMAIC in the toolbox	Fewer tools in the toolbox—different purpose
Can be the DNA, culture, philosophy, thinking and standard of excellence of a company	Can be the philosophy and thinking of a company
Can be achieved without Lean	Provides tools to sustain Six Sigma

FIGURE 10.3. Comparison of Six Sigma and Lean.

health care have all been cited from time to time as having implemented Lean concepts.

Six Sigma implementation has further strengthened the viability of Lean tools in creating lasting solutions. Organizations applying Lean strategy to improve productivity, quality, and profitability have consistently cited quantum improvements in many process areas. Regardless of the size of an organization, significant and unprecedented gains and improvements are typical with any successful lean initiative or implementation:

Cycle time	Reduced 50 to 60 percent
WIP inventory	Reduced 60 to 80 percent
Productivity	Increased 50 to 120 percent
Lead time	Reduced 40 to 80 percent

DMAIC PHASE	KEY SIX SIGMA TOOLS	KEY LEAN TOOLS	ADDITIONAL BENEFITS OF INTEGRATING LEAN AND SIX SIGMA
Define	Kano analysis, Pareto analysis, process mapping, SIPOC model	Value-stream mapping (VSM)	Problem clarity and improved customer perspective
Measure	Data normality test, COPQ, MSA, C_p, C_{pk}, DPU, DPMO	Cycle time, muda, yield	Sources of non-value-added activity
Analyze	Root-cause analysis, MVA, FMEA, hypothesis, ANOVA	Constraints, TAKT time, cycle time	Better cause and effect relationships in process and product characteristics
Improve	TRIZ, component search, factorial designs, RSM	VSM future state, continuous flow, kaizen, 5-S	Efficient process flow design with capable processes
Control	Control charts, precontrol charts, process model	5-S, TPM, visual controls, kanban	Highly efficient and effective workplace; sustained customer focus

FIGURE 10.4.   Six Sigma and Lean Tools in DMAIC methodology.

| Throughput | Increased 40 to 80 percent |
| Space requirements | Reduced 20 to 40 percent |

Additional benefits that are visible on the floor include better housekeeping, more organized workplace, more open space, higher inventory turns, and more knowledgeable and motivated employees. Inventory is considered central to Lean efforts. Inventory represents materials not in use, hence creating a loss or waste. The inventory buildup might have occurred as a result of current accounting practices, batch processes, economical purchase quantities, or sheer practice of producing the product as long as the process is running well. Under Lean thinking, it is waste, nevertheless.

In the context of Six Sigma, Lean integration provides standardization and discipline, a flow mentality, and tools for identifying and eliminating non-value-added activities. An ideal integration will produce results that are far superior to those of any one approach alone.

PRINCIPLES OF LEAN THINKING

The management at Toyota has documented its philosophy in its TPS publications. The purpose of TPS is to create a system for conserving resources by eliminating waste. TPS prepares people to recognize waste and take initiatives to eliminate it and prevent its recurrence. The leadership treats its facilities as a showcase for customers and communities to see and to demonstrate how it helps employees to master new skills and take new initiatives. TPS bases its strength in the employees' creativity and empowerment to continually improve the workplace and increase work fulfillment. Lasting gains in productivity and quality are possible when management and employees are united in a commitment to positive change. TPS gives people a chance to become as good as they can be.

TPS is a trust-based system in which employees and management develop trust in each other and remove barriers to improvement. Gains made through improvement activities and productivity increases are shared with those contributing to the success. At Toyota, employee satisfaction is the key to customer satisfaction. TPS has two basic goals—customer satisfaction and corporate vitality.

Lean manufacturing also means waste-free production. Waste-free production means producing what is needed, when it is needed, and at a specific rate of production. Meeting customer demand is more important than keeping assets busy. Once customer requirements are determined from a Lean perspective, resources and plans are prepared to achieve those objectives. In the Lean system, any changes, abnormalities, or fluctuations in the flow of materials, the operation of machines, or the environment become easily visible and call for immediate action. Any unplanned changes in a Lean environment hamper the capability of a Lean system to deliver product on an as-needed basis. It is for this reason that Lean systems are required to have very high reliability, consistency, and standardization.

Lean is an evolving process, developed by human ingenuity and creativity. Lean thinking is a Western adaptation of TPS. Lean concepts remain essentially unchanged from those in TPS. However, the processes and beliefs may differ, reflecting the effects of differing cultures. TPS develops a customer focus by emphasizing employee satisfaction through creativity, development, empowerment, and recognition processes. Lean thinking develops customer focus by taking a process approach, the human element being part of the process. Figure 10.5 shows the key Lean tools in the toolbox.

Lean combined with Six Sigma uses a set of concepts or beliefs that form the foundation for sustained dramatic improvement. While all these tools are very powerful, we shall focus and elaborate on only those tools that have very specific relevance in Six Sigma. These tools complement the efforts of Six Sigma and when applied in combination with Six Sigma tools provide a significant value addition to the initiative effort.

LEAN TOOLS FOR SIX SIGMA

The Lean tools may be applied in various phases of the Six Sigma DMAIC method. The tools that have the most signifi-

5-S	Total productive maintenance (TPM)
Five *whys*	Work balancing
Visual factory	Cell
Muda	Single-piece flow
Poka-yoke	Kanban
Single-minute exchange of dies	Value-stream mapping (VSM)

FIGURE 10.5.   Lean tools.

cant effect and their placement in the methodology are shown in Fig. 10.6. They were selected due to their relevance to the phases of Six Sigma, as described in the following subsections:

Value-Stream Mapping (VSM)

Value-stream mapping (VSM) is useful in all phases of the DMAIC methodology. In the Define phase, VSM provides a clear view of the current state or as-is process flow. It is possible that in looking at the current state, the problem not only will be further clarified, but sometimes will become so obvious that it can be eliminated without the further application of any other sophisticated tool.

In the Measure phase, the necessary process measurements are made. In the Analyze phase, all the elements are examined for their relevance to value addition. In the Improve phase, a future-state value-stream map is developed to allow review of the effects of planned changes in the process. With improved capabilities (reduced variation through Six Sigma), it is possible that the process flow will add more value and contain less waste. The future-state value-stream map will reflect knowledge-based change and will be more efficient as a result. In the Control phase, the necessary controls are established to ensure that the improvements sought are actually sustained.

The major benefits of using VSM are as follows:

- Provides an excellent visual display of the big and relevant picture

Phase	Applicable Lean Tools
Define	VSM, muda, 5-S
Measure	VSM, TPM
Analyze	VSM, TPM
Improve	VSM, 5-S, TPM
Control	5-S, muda, TPM

FIGURE 10.6. Selected Lean tools in DMAIC methodology.

- Links upstream and downstream processes effectively
- Displays both the information and material flows with their dimensions
- Provides a complete understanding of the major elements of the product cost
- Helps to identify the process bottlenecks
- Not only helps to identify the waste but also points to its source
- Makes waste reduction opportunities too obvious to ignore
- Helps to prioritize the efforts by identifying the biggest opportunities for improvement
- Becomes the foundation for implementing the necessary change
- Sets the common language within the organization for executing the improvements
- Becomes the foundation for continuous improvements

All the preceding benefits are available to organizations that plan to transform from traditional to Lean enterprises.

VSM is an *end-to-end* view of the process flow (stream) that creates customer value for a given product or service. It follows a product's information path from customer to supplier through the organization, then follows the product's production path back from supplier to customer, through the organization. It includes all the actions, those that are adding value for the customer and those that are not adding value for the customer. It provides the big picture of the process and an opportunity to see where waste exists. The symbols that are commonly used in VSM are depicted in Fig. 10.7.

A sample value-stream map is presented in Fig. 10.8. In this value map, the customer (top right corner of figure) places orders through multiple modes at an agreed schedule frequency. The production planning department translates each order into many actions, including ordering from its supplier (top left corner of figure) and scheduling the operations within the organization. The supplier delivers the ordered material to

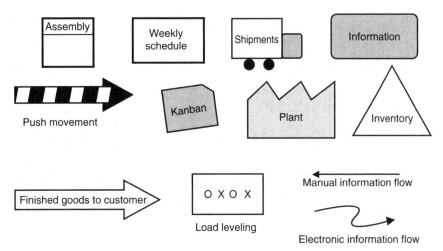

FIGURE 10.7.   Partial list of value-stream mapping symbols.

the organization, where it waits (lower left corner of figure) for some time (unit is in days). Operation 1 picks it up and processes it; this normally takes a much shorter time, compared to the waiting time before the operation. After the first operation, the material again waits for a considerable time before the second operation is performed. Here we assume that the material passes through only two operations before it is completely processed for shipment. After the second operation, it is ready for shipment and is shipped per customer demand.

Looking at this value-stream map, note that a significant amount of time is spent in waiting and only a small portion of the time is actually spent in the processing of the material. The production lead time and the processing time are defined at the bottom right corner of the figure. The ratio of processing time to production lead time is the indicator of how lean the system is. In a typical processing environment, it is a very small number.

Value-stream mapping is the same as process mapping, except that it also includes relevant process information such as number of resources, inventory, output requirements, and cycle time to complete the process. It is different from the process map, commonly used in Six Sigma methodology, in the following ways:

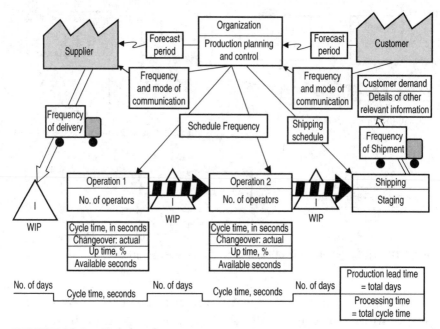

FIGURE 10.8.   Sample value-stream map.

- It provides a broader picture of the process with higher-level details.
- It provides more information in terms of resources, time required to complete the activity, and other process dimensions.
- It facilitates the identification of potential opportunities for improvement by providing this additional information.

VSM serves as a starting point to help project leaders recognize and identify the sources of waste in the process through its pictorial display of activities. It serves as the means to prioritize improvement activities and flow their benefits through to the bottom line. The generic steps in achieving improvements through VSM are detailed in Fig. 10.9. VSM activities should be carried out by a team of middle- and upper-management personnel, because of their familiarity with scheduling, planning, and other information.

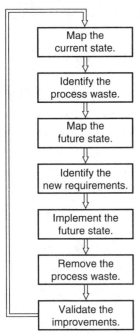

FIGURE 10.9.   Value-stream improvement.

You use a map to reach a desired destination, and the value-stream map is no different. It leads you on the path of future improvement. The steps in achieving that improvement are briefly described in the following paragraphs.

1. *Map the current state.*   This is the mapping of the *as-is* process performance. It is a walk through the target process from beginning to end, observing each operation along the way. This gives a general idea of areas to cover, things to document, value-added mindset, and so forth. Some consultants suggest that VSM should start from the end of the process, that is, proceed upstream. This helps to clarify how processes link in the map and introduces the concept of pushing versus pulling. It is important to include material flow as well as information flow on the same map. Usually, it is a good idea to have all team members draw the current-state VSM independently of each other, allowing review of the process from different perspectives. Later all information from the team members can be con-

solidated into one VSM. This also revalidates all information collected by the individual team members.

2. *Identify the process waste.*   This is the analysis part of VSM, and this part delivers the true objective of using the Lean system. Identify value-adding and non-value-adding activities. Think of necessary versus unnecessary, planned versus unplanned activities. What can be changed to optimize the process? This step identifies all possible waste. (The waste element is discussed in the next subsection, Muda.)

3. *Map the future state.*   This step defines the future state of the process, when all the identified waste is assumed to have been removed. This expresses the expected improvement of the exercise initiated by VSM. The future state sometimes goes through several revisions before it is ready for further action.

4. *Identify the new requirements.*   The proposed waste removal will create major changes in the process, and this will generate the new training needs in the organization. Before successfully implementing the required change, it is important to understand all the implications and communicate with all affected personnel and value-chain partners.

5. *Implement the future state.*   Prepare for implementation by providing necessary training to the personnel and communicating with the partners of the value chain. Once complete, prepare a plan for the implementation of the future state. This is when the knowledge of Lean tools such as kanban, kaizen, TPM, 5-S, and others is required in order to achieve the desired quality, efficiencies, and flow. A subject-matter expert may be utilized as a mentor and guide through the first few implementations.

6. *Remove the process waste.*   This is the actual action part of VSM. The identified waste is removed from the process in a planned manner.

7. *Validate the improvements.*   After the future state has been implemented, it is important to have the stated benefits validated by independent auditors. This not only boosts confidence in the process but also sets the stage for further improvements by continuing the cycle.

It is important to note that the preceding seven steps are iterative in nature. With the dramatically changing business environment (i.e., ever-demanding customers and continuous improvements in technology), opportunities for waste reduction may be available on regular basis. It is therefore important to continue this exercise on a periodic basis unless some external event demands it be done faster.

MUDA

Muda is a Japanese term that means waste or wasteful activities. Muda is a defect, an undesirable condition, any activity that does not add value. *Muda, waste,* and *non-value-adding* are interchangeable terms. A non-value-adding activity is one that the customer does not pay for, but which the organization performs to suit its own production system, practice, beliefs, or culture. Situations such as inventory stored on the shelf, WIP, parts waiting for the next operation, a broken or dull drill, or a shortage of printing paper all signal the presence of muda in the system.

Waste elimination is the lowest-hanging fruit in any organization, and it has a direct bearing on the bottom line. Any process executed in the organization either produces a value that the customer is willing to pay for or just produces something that the customer does not desire. Anything the customer does not desire or pay for is a waste. Some organizations tend to be so obsessed with their internal focus that they forget that the customers exist, and such organizations are in fact the biggest producers of waste.

In the Measure phase, Six Sigma quantifies product defects in terms of DPU, DPMO, and so forth, while Lean identifies process muda. Together they provide more comprehensive opportunities for solid improvements in both product and process quality. Muda is relevant to all phases of the Six Sigma DMAIC method. After all, the end objective of Six Sigma projects is to improve the bottom line by waste elimination and full customer satisfaction. You identify the waste in the Define phase, measure it in the Measure phase, and so on until you ensure that the controls are in place to keep waste at its minimum level.

Taiichi Ohno identified seven types of waste or muda. Application of Lean thinking can correct and eliminate much of the waste in any system. Following are detailed descriptions of all the muda originally identified.

Muda 1: Overproduction.   When a company or department produces more than what is really needed to meet customer demand, the excess quantity is called *overproduction*. Building to stock (forecast), producing more to take advantage of a good setup, finishing the material left over from a run, running in fixed lots, or compensating for defects are examples of overproduction. This is also called *just-in-case* production. The key waste elements from this source include the following:

- Disruption to the smooth flow of material or information
- Increased lead time
- High storage costs
- Complicated defect handling
- Unknown problems in the system (most other inefficiencies are hidden under cover of the overproduction, such as improper functioning of the inspection process)

Improving setup reliability and making faster and more frequent setups can correct the problem of overproduction. Producing only what the customer requires when it is required improves quality and productivity. This requires a streamlined, well-oiled supply chain and the participation of all the process players.

One way to keep the supply chain streamlined may be to minimize the shocks that lead to the bullwhip effect. For example, Wal-Mart pursues a policy of always having low prices. This means that the customers expect low prices every day, and their expectation level is set accordingly. The majority of purchases from Wal-Mart would thus be per the actual needs of the customers, not strategic purchases because of some special run by the store. The demand pattern in this scenario would be much more stable and predictable than the major fluctuations created with the deals, promotions, sales, and similar tactics commonly pursued by retail organizations.

Muda 2: Transportation. Moving products and materials from one process to another doesn't add any value. Moving product to inventory and then bringing it back to the line is one such example. Non-value-adding transportation results from overproduction, lot (batch) sizing, inefficient plant layouts, and availability of large storage areas, to name a few causes. The major losses caused by this waste include the following:

- Risk of material damage
- Cost of the additional transport equipment and personnel

Utilizing a pull system, kanban, or flow mentality minimizes unneeded transportation to and from the workstation.

Muda 3: Motion. In this case, *motion* refers to *human* motion. It focuses on the ergonomics of the workplace. When the workstation is not laid out to facilitate natural human movements, a stress is created. Examples of wasted human motion include reaching for parts (ergonomics), tiring actions (work-cell design), searching for parts and tools, walking long distances to fetch parts or tools, and the like. Muda of motion affects physical conditioning, attitude, and motivation. The direct losses from this waste include the following:

- Excessive fatigue, resulting in reduced physical and mental productivity
- Increased absenteeism due to illness or fatigue

Some Lean tools that can eliminate such muda are 5-S, visual systems, and kaizen.

Muda 4: Waiting. Waiting involves idle time due to unscheduled machine stoppages, personnel, parts or material, or material unavailability and quality issues. If one of the inputs or outputs is not ready to continue for any reason, the other elements wait for it to become available, creating muda. Examples include situations when an operator waits for first-piece inspection or setup approval, or when an invoice from the boss has not been approved for 5 days. Waiting is mostly unproductive and adversely affects total output and cycle time. The major costs associated with this waste include the following:

- Reduced cash flow due to the blockage of cash for more production than required
- Reduced opportunity to take the benefits of short-term cash investments
- Risk of damage in the case of perishable items or obsolescence in the case of high-technology products

Activities and initiatives such as planning production to TAKT time, installing in-process measurements, total productive maintenance, and kaizen help to minimize waiting.

Muda 5: Processing. Process operations that the customer is not willing to pay for add processing muda. Examples include multiple inspections, paperwork problems, extra cleaning operations, added deburring, and the like. Such actions are normally the result of inadequate product or process design, production planning, material problems, or machine capabilities. The major losses associated with this waste include the following:

- Excessive cash tied up in overdesigned, expensive processing equipment
- Excessive personnel working at cross purposes, reducing the overall effectiveness

Establishing a flow kaizen and office kaizen and creating a pull system can address many of the issues related to process muda.

Muda 6: Inventory. Inventory is a psychological safe haven for many organizations. Inventory is built by design as well as by chance through overproduction. Inventory is also perhaps the biggest "bank" that provides little return on early investment. Inventory generally falls in the big-muda category. Inventory is a waste that lives in various forms, such as finished goods, work in process, supplies, purchased products, and raw materials. Higher inventory levels are synonymous with greater incentive for push-type systems. Higher inventory means lower inventory turns and possibly a higher breakeven point. Inventory buildup may happen due to lead-time variations, long setups, long process cycles, purchasing practices, and economic batch sizing. The losses associated with this waste include the following:

- Obsolescence due to customer requirement shifts or technological changes
- Scrap due to expiration of the shelf life, particularly in the chemical and pharmaceutical industries
- Blockage of the expensive floor space
- Carrying costs, including insurance, care, and opportunity costs
- Delays in problem identification, risking the permanent loss of the customer

Lean tools such as kanban, flow processes, and short setups tend to provide reduction in inventory levels and implementation of flow mentality. Also, it is vital to develop a strategic partnership with the suppliers to improve their reliability to supply the products and services on time. The WIP may be reduced by improving machine reliability. Lean provides a very powerful tool to improve this reliability—total productive maintenance, which is discussed later in this chapter.

Muda 7: Defects. Any error or lack of something necessary in the system creates muda in the form of a defect, such as scrap, rework, field failure, excess variation, and missing parts. Defect muda may be caused by failures in the system, batch processing, inspection failures, and machine capability issues. The losses associated with defects include the following:

- Direct loss of revenue, that is, an adverse effect on the bottom line
- Loss of goodwill in the marketplace if shipped to the customer

Establishing a pull system, flow manufacturing, and building in quality provide good countermeasures to address defect muda.

In addition to the preceding seven mudas, many companies have added one more type of waste to the list—*underutilization of human resources.* This means ineffective utilization of human energy and mental power. This muda happens when people are not properly trained, skilled, or equipped to do their job effectively; resources are not planned according to the demand requirements and process capabilities; and work is not designed in a way that is conducive to producing quality. The major losses associated with underutilization of human resources include the following:

- Missing true opportunities to achieve operational excellence. After all, the person who wears the shoe knows where it pinches. Employees are the true source of ideas.
- Missing opportunities to achieve higher levels of human performance due to the absence of potential motivational opportunities that would have existed with the active participation of employees.

A good practice is to carry a checklist of the seven wastes from work center to work center and identify all the waste one can find. Often it is not possible to differentiate between waste and value added due to our limited understanding of the concept. Again, having team members go out and identify waste first independently and then collectively will help to identify most of the muda activities. About 50 percent of all muda activities are found to be unnecessary and can be promptly eliminated, creating a more streamlined and leaner future-state value-stream map.

5-S

Six Sigma requires dramatic improvement in a process—by a factor of more than 20,000 when improving from Three Sigma (66,810 ppm) to Six Sigma (3.4 ppm). An improvement of such magnitude requires everyone's involvement. 5-S is a great housekeeping method that gets employees involved in keeping the workplace clean and organized. Employees start to think about improving their processes. The adoption of 5-S from Lean tools to Six Sigma methodology is an example of continual improvement of the Six Sigma methodology.

The 5-S method has been used in cell manufacturing, where the organization of cell, flow, and other factors prevent the presence of any waste- or delay-causing activities. In order to reap greater rewards from the 5-S process, it must be part of a larger strategy like Six Sigma, where followthrough and sustainability of effort and achievement are given serious thought.

5-S starts right from the Define phase of the Six Sigma DMAIC method. After all, any improvement initiative can be successful only with neat housekeeping. 5-S provides added support to Six Sigma plan in the Improve and Control phases. While Six Sigma determines the best solution to the problem, 5-S provides tools for maintaining the improved state of workplace organization.

5-S provides five pillars of workplace organization. 5-S is also known as a housekeeping program, and housekeeping is a prerequisite for process improvement. Sustaining any improvement would be difficult without proper workplace organization. The 5-S method is represented by five Japanese words that begin with the letter S in the Latin alphabet. Their English translations are given in parentheses in the following descriptions.

1. *Seiko (sort).*   Clear out everything; sort and eliminate what is not needed. This means segregate what is needed on the basis of frequency of usage. The items that are needed frequently must be kept near the workstation, and those needed infrequently must be kept in the storage area. Those that are not needed must be disposed of economically. A properly sorted

workplace reduces the probability of making defective products by using the wrong materials or tool.

2. *Seiton (straighten).*   Create a place for everything, and put everything in its place. Once the items around the workplace have been sorted, the places to store them must be well defined. This not only helps operators find the required items more quickly but also eliminates the possibility of creating waste by reordering what is already available in stock. Identify and define locations based on frequency of use, ease of storage, and so forth, with the most frequently used items placed closest to the workstation.

3. *Seiketso (shine).*   Clean everything. The workplace should be spotless. This is the key feature of the *Lean and Clean* system. The workplace must always be ready to welcome a customer. A clean work environment not only increases customer confidence but also keeps employees motivated and happy to perform their activities effectively and efficiently. This new look of the workplace not only improves productivity but also improves safety. Remove all buildup, oils, and dirt in and around the workstation. Fix sources of such contamination. Fix leaky machines or contain the leaks. Paint machines to make their appearance pleasant, as in a showroom.

4. *Seiso (standardize).*   Once the workplace is clean, that cleanliness needs to be sustained. This means that there must be a system to ensure that the best practices are followed. This can be accomplished only when the process has been standardized and the checks and cleaning are performed regularly.

5. *Shitsuke (sustain).*   This step is the true foundation of the 5-S method. Without any effort to sustain the improvements, the system may slip back to its old undesirable state. Therefore, it is important to sustain the standardized system and make it a way of life. Train and maintain discipline. This aligns practices to comply with the plan and creates a no-excuses environment. The way to do this is through periodic surprise internal audits.

There is a need to allocate time for 5-S in every activity. The interrelationship of 5-S is pictorially represented in Fig. 10.10.

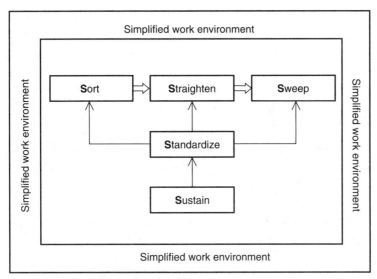

FIGURE 10.10. 5-S interrelationship.

The 5-S approach is based on the philosophy that a clean
and organized workplace leads to greater willingness to do
things right, leading in turn to higher productivity and quality.
Conversely, a cluttered workplace is not a productive workplace. The Japanese describe it as the *showroom mentality*. A
good showroom, where people have positive attitudes, not only
attracts lots of visitors, but also conveys a feeling of confidence
and trust. By treating one's workplace as a showroom, the
employees also feel positive about their work, quality, and productivity. Dirt cannot hide in a 5-S workplace. It is emphasized
that the cleaning must be done by those who work at the workstation. The intent is that by doing the cleaning, employees
become more vigilant and in touch with daily changes in
machine function, sources of new problems, leaks, deterioration, performance, and noise and can take immediate action to
fix the source of the problem.

Sometimes it is may not make economic sense to fix a problem, such as an oil leak in an old machine for which the cost
of replacement parts is prohibitive. In such cases, the employees plan actions to contain the problem, such as neatly placing

a drip bucket, which is then emptied when full. The financial loss is minimized without affecting the success of 5-S.

Walking around one's workplace it is easy to detect items that are not needed, such as old papers; dirt; chips; clutter under the machines; leaks; mist; spots on the machines; chipped paint, tools, and fixtures; broken pencils; and many more things. Such items can be noticed in the corners and aisles; under the desks, in drawers, on machines, cabinets, and walls; and so forth. A trained eye and an understanding of what to look for in making the product and process a success make the 5-S process a greater success.

Achievement of the 5-S goals requires the commitment and involvement of leadership. Well-intended 5-S efforts often fail when the sense of purpose is lost and it is treated like just another program—especially in our mentality, when we choose to ignore or postpone the important (e.g., the showroom mentality) in the presence of the urgent (e.g., late orders).

TOTAL PRODUCTIVE MAINTENANCE (TPM)

In manufacturing operations, it is critical to maintain optimal machine conditions to minimize variance. When implementing Six Sigma in an equipment-based manufacturing or service environment, a good maintenance program must also be implemented to ensure reproducibility. The objective of total productive maintenance (TPM) is to develop optimal machine conditions that will also extend the economic life of the machine.

Conventional maintenance programs believed that maintenance needed to be done only when the machine broke down (i.e., it would no longer run). This type of maintenance was also called *breakdown* or *reactive maintenance*. It accepted breakdowns as normal in a machine's service life and a necessary cost of doing business. This situation obviously resulted in the unpredictable loss of time and the failure to make the commitment to meet the customer requirements on time. As competition increased, both machine users and machine manufacturers had to rethink the maintenance strategy to overcome this handicap. Machine manufacturers then started to share information about keeping the machines operational and started to recommend periodic care. This included changing certain parts and replacing consumables on predefined schedule. This theory of maintenance came to be known as *preventive maintenance*. It no doubt reduced machine breakdown, but it became uneconomical to sustain due to high costs associated with the spare parts and the labor involved in executing every recommendation.

Further intense competition forced organizations to think differently and more scientifically. The engineers started to focus on the reliability of the machines by equipping the team to handle the maintenance issues. The new concept came to be known as *total productive maintenance*, but in reality it is total *participation* maintenance. TPM combines preventive maintenance with quality control and total employee involvement. The key objective is to minimize the machine downtime caused by maintenance efforts and sustain peak machine performance at all times. It covers the entire life cycle of the machine, and its implementation focuses on the following points:

- Teamwork, with the operator responsible for the machine's condition—not an "I break it and you fix it" approach
- Training of operators in general machine care such as daily checks, lubrication, part replacement, repairs, precision checks, and early detection of abnormal conditions
- Development of necessary standards for machine operation and monitoring
- Improvement in machine design for increased reliability
- Elimination of breakdowns
- Elimination of adjustments
- Elimination of product defects due to machine condition

The scope of TPM covers preventive maintenance, breakdown maintenance, equipment improvement (through equipment kaizen), and maintenance prevention. Successful implementation of TPM can achieve reductions of more than 90 percent in the number of breakdowns and defect and repair rates. Cleaning time and the consumption of hydraulic fluids are also reduced by a significant amount.

TPM is a powerful tool for achieving higher machine uptime and capability. The objectives of TPM are to create optimal machine conditions and to improve overall machine conditions. In order to achieve TPM perfection, there should be zero machine adjustments, zero machine breakdowns, and zero machine defects. TPM supports the kaizen philosophy by giving control of and responsibility for the improvement of certain machine maintenance activities directly to operators, team members, and supervisors. In essence, TPM seeks to improve the performance of the equipment rather than just maintain the status quo. The TPM and conventional maintenance approaches are compared in Figs. 10.11 and 10.12.

Therefore, TPM works on the belief that targets of zero defects and zero breakdowns are possible and achievable. The classic work system and machine maintenance program create certain losses and deprive the operator of full productivity. Figure 10.13 describes these losses, which are potential opportunities for improvement.

The implementation of TPM also increases workplace efficiency and is an integral part of the Lean approach. In the Six Sigma Control phase, TPM supports the sustainability of problem solutions. TPM creates optimal machine conditions through human–machine interaction or autonomous maintenance. It brings the following preventive maintenance mindset:

- Recognize even the smallest abnormalities in equipment, tools, and fixtures.
- Believe that no problem is too small.
- Establish the habit of cleaning and checking.
- Change the way you see and think about the things around you.

TPM uses 5-S and kaizen activities to improve the performance of equipment and eliminate the need for frequent main-

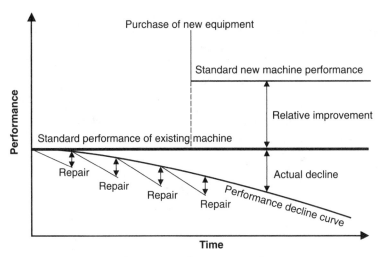

FIGURE 10.11. Performance with conventional maintenance.

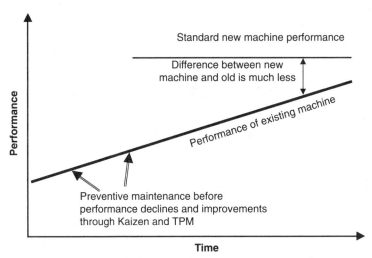

FIGURE 10.12. Performance with TPM.

tenance. Equipment cleaning by the operator is a major part of TPM. Cleaning by hand offers the following philosophical advantages:

- Cleaning causes the operator to feel, look, and check.
- Checking leads to the discovery of minor problems.
- Minor problems are fixed or improved.

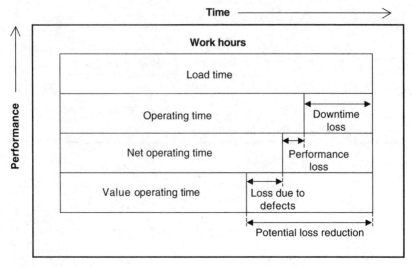

FIGURE 10.13. Opportunities for improvement: losses due to breakdown, setup and adjustment, idling and minor stoppage, reduced speed, quality defects and rework, startup, and reduced yield.

- Fixing problems when they are still small produces a positive environment.
- A positive environment produces greater motivation.

TPM is a human–machine interaction in which operators play a pivotal role in creating optimal machine conditions. A TPM-oriented system prepares operators to recognize even the smallest abnormalities and defects in machines, tools, and fixtures. Operators and maintenance work as a team to correct the problem immediately and eliminate the cause to prevent recurrence. TPM helps establish the practice of clean–check–restore–improve to ensure that people change the way they think about things around them.

TPM identifies six major losses from each operation:

1. Losses due to breakdown
2. Losses due to setup and adjustment

3. Losses due to idling and minor stoppage
4. Losses due to reduced speed
5. Losses due to quality defects and rework
6. Losses due to startup and reduced yields

These losses are considered obstacles to the achievement of higher efficiency and must therefore be eliminated.

FREQUENTLY ASKED QUESTIONS

SIX SIGMA BASICS

Why is quality important?

Quality has a direct effect on market share and the total cost of the product. Good quality is important in every aspect of product competitiveness—for example, to retain or increase market share, charge premium prices, improve customer satisfaction, and sell higher volumes. Among other things, poor quality results in higher costs, generally referred to as the *cost of quality* or *cost of poor quality.* The cost of quality can be categorized as *prevention* cost, *appraisal* cost, *internal failure* cost, and *external failure* cost. On the other hand, good quality leads to higher customer satisfaction, lower rework cost, fewer rejects, lower warranty costs, and higher productivity (due to lower variation in the system.) It is thus fairly easy to see that good quality is a key to the sound financial health and longevity of a company.

What is total quality?

Total quality encompasses the whole organization and stresses a continuous commitment to excellence in all aspects of company operations. It spans across the organization, including the suppliers. The key to total quality is the involvement of everyone in delivering quality products or services to the end customer. Quality is everyone's Job 1. In order to achieve the goal of good quality, effective procedures should be instituted to support employees in building quality products, and the employees should be empowered to modify systems and procedures as required—for example, through quality circles. Since quality is defined by the customer, the overarching objective of total quality is to exceed the customer's needs and expectations.

Why are engineering processes important for business success?

The effectiveness of a company's engineering processes determines its productivity. There is a strong correlation between the quality of the product and the reliability of the installed processes. In the current competitive environment, effective processes also allow the company to achieve flexibility and speed, so critical to its very survival.

What is process management?

Effective processes at every level are critical for an organization to execute its strategy. A strategy lays out the path for the organization to accomplish its goals, defined in its mission, which in turn is driven by its vision. As a result, managing processes at all levels is a critical component of a company's ability to meet the customer's expectations. Process management thus enables all activities in the organization. The following are key components of effective process management:

- Defining output requirements
- Identifying standard work
- Instituting procedures for monitoring the process
- Reviewing and providing immediate feedback
- Taking corrective actions, if required

What is statistical process control (SPC), and why is it important?

The use of statistical methods such as control charts and capability analysis to monitor and control a process is collectively referred to as *statistical process control* (SPC). SPC assists users in identifying statistically out-of-control processes so that immediate action can be taken to bring them under control. A process is said to be in *statistical control* when only natural variation or random noise remains in the process and all special causes of variation have been eliminated. Using statistical tools, the presence of unwanted variation is illustrated by the data that fall outside of predefined control limits on the control charts. SPC methods can be applied to any type of process, which can be either a service or a manufacturing unit.

SPC is important because it can prevent defects, thus *building* quality into the product, rather than *inspecting* quality into the product. Needless to say, preventing defects is a more effective strategy. Once again, we highlight the relationship between process reliability and quality of output. SPC methods can also help in predicting the future performance of the process using historical information and in proactively recommending corrective actions, if necessary.

What are ISO 9000 and QS-9000 standards?

The ISO 9000 standards were created in the 1980s by a consortium of 12 European nations. The ISO 9000 standards serve as a universal quality benchmark for companies worldwide. These standards are basic quality standards that apply to a broad range of industrial and economic sectors. The standards offer broad guidance for quality management and general requirements for quality assurance. However, these standards describe various items to be included in the quality system, without discussing the methodology for implementing such a scheme. Needless to say, unless building quality products is at the core of the company's philosophy, quality initiatives are mere slogans and programs that come with an expiration date. As a result, the actual design and implementation of any quality system depends on many factors, including the company's core values, products, processes, choice of technology, resources, and market environment.

The QS-9000 standards were established by the Automotive Industry Action Group (AIAG), following in the footsteps of the ISO 9000 standards. The big three automakers in the United States as well as most large truck makers were involved in the development of these standards. Like the ISO

9000 standards, the QS-9000 standards provide a broad guideline that includes all elements of the ISO 9000 standards plus
special requirements for the automotive and truck industry.
The QS-9000 standards replaced all previous automotive supplier quality programs.

What does the term *Six Sigma* mean?
In statistics, the Greek letter sigma (σ) represents the standard
deviation (SD) of a population. In other words, σ characterizes
the variability about the average value of the data representing
the population. A large σ value implies a bigger difference
between the maximum and the minimum values in the data
set. Furthermore, the underlying assumption of Six Sigma
methodology is that all data follows a normal distribution.
Consequent to the properties of the normal distribution, about
68 percent of the data values will fall within 1 σ of the average
[68 percent of values will fall between (average value $-\sigma$) and
(average value $+\sigma$)]. Similarly, about 95 percent of values will
fall within 2 σ, 99.7 percent will fall within 3 σ, and so on. A
σ level of 6 (or 6 SD) for a process implies an average of only
3.4 defects in 1 million opportunities.

What is the goal of Six Sigma?
The main goal of Six Sigma is to deliver significant real improvements that are reflected in the improved overall bottom-line
financial numbers through continuous process improvements.

What is the defect level of a Six Sigma process?
In a true Six Sigma process, only 3.4 defects are present in 1 million opportunities (also referred to as 3.4 ppm.) However, due to
practical issues and limitations, most good companies today
operate at the level of Three Sigma to Four Sigma (defect rate of
67,000 to 6,200 ppm.) It should be noted that the presence of
these undesired defects, however small, in the output results in
additional costs—described earlier as the *cost of quality.*

What is the Six Sigma methodology?
The Six Sigma methodology originated at Motorola in the
1980s. Essentially, it was deployed to monitor and control the
process variation. Before discussing the methodology itself, it

is important to note that "Six Sigma" is more than just a technique or a method. The meaning of Six Sigma depends on various factors, such as the objective, the scope, and the expected results. Accordingly, Six Sigma can be a philosophy for achieving business improvement, a structured approach for reaching an improvement goal, or a measurement of performance for comparing and monitoring business processes (sigma score). The Six Sigma methodology has been in use for over 20 years and successful in delivering value where business decisions are based on facts and data. The underlying principle of the Six Sigma methodology is to identify the root causes of a defect. Specifically, Six Sigma operates by the determination of actions through examination, analysis, and experimentation of the process along with the level of performance required to produce quality output. Essentially, a Six Sigma program drives and manages the process using statistical controls to ensure consistently good quality of the process and hence of the resulting output. The value of sigma plays a key role in the design and implementation of such a program.

Why focus on improving sigma?

The goal of any operation is to consistently deliver a good finished product to the customer. The root cause of all problems related to consistency and quality is the presence of controllable and uncontrollable (natural) process variations in different forms. The key focus of the Six Sigma methodology is the gradual reduction of the overall process variation. By improving the sigma value, the variation is brought under control, leading to typical benefits, such as the following:

- Lower overall costs
- Higher resource utilization
- Higher employee morale
- Higher customer service levels
- Higher profitability

What are the benefits of Six Sigma?

The most commonly realized benefits of Six Sigma are the following:

- Higher productivity
- Reduction in cycle time
- Higher throughput
- Superior quality and reduction in defects
- Standardized improvement methodology across the organization
- Standardized sets of tools and techniques to simplify improvement efforts
- Quality consciousness as a part of the work culture
- Higher customer satisfaction levels
- Significant improvement in the bottom line

What are the key elements of the Six Sigma methodology?
The key elements of the Six Sigma methodology are the following:

- Data- and fact-driven decision-making approach
- Use of simple statistical tools to analyze process data
- Standardized process improvement methodology
- Effective measurement systems
- Standardized metrics to drive improvement methodology across an organization
- Use of design of experiments (DOE) to optimize process performance

What is DMAIC?
DMAIC is a methodology for process improvement within the Six Sigma framework. The DMAIC phases are as follows:

- *Define.*   Define the project, process, and customer requirements.
- *Measure.*   Measure the current process performance.
- *Analyze.*   Analyze data to determine and verify the root cause of the problem.

- *Improve.* Implement solutions that address the root causes.
- *Control.* Maintain the gains.

It is a systematic process improvement methodology that ensures that the root cause of process variation is identified and addressed in such a way that the process is permanently improved.

What is DMADV?
DMADV is used to design a process, product, or service. The DMADV phases are as follows:

- *Define.* Define the project.
- *Measure.* Measure and determine customer needs.
- *Analyze.* Analyze the options to select one that best meets customer needs.
- *Design.* Design the detailed processes.
- *Verify.* Verify the design performance.

This systematic design methodology can build processes that are very reliable in meeting customer expectations.

Which methodology is better: DMAIC or DMADV?
The similarities of DMAIC and DMADV are as follows:

- Both are Six Sigma methodologies.
- Both are data- and fact-based solution approaches.
- Both are ways to help improve business processes.
- Both lead to bottom-line financial improvements.
- Both are implemented by Green Belts, Black Belts, and Master Black Belts.
- Both need the support of a champion and process owner to be successful.

The differences of DMAIC and DMADV are as follows:

DMAIC methodology is used when:
  - A product or process is in existence but is not meeting customer specifications.

- The product or process is not performing adequately.

DMADV methodology is used when:
  - A product or process is not in existence.
  - The existing product or process has been optimized and still doesn't meet expectations.

A common mistake encountered in Six Sigma projects is the choice of the incorrect methodology. Often a project is scoped as needing DMAIC for incremental process improvement when it really requires a DMADV methodology improvement. Obviously, in such an event, the gains are limited. To fix this problem, every effort should be made to use the knowledge gained from the failed exercise to launch a well-designed DMADV project.

How does Six Sigma help me focus my business operations?
Among other things, the Six Sigma methodology brings an effective structure to process management. Since the focus is on the customer's definition of quality, the Six Sigma project demands identification of the business strategy and its key elements for success. In addition, it requires the identification and analysis of gaps that may lead to failure. This process helps in the selection and prioritization of projects that will truly add value to the end product in the customer's view. Iteration through this rigorous methodology based on extensive data collection and analysis helps in the identification of the real root causes, which affect process performance, and also allows the development and implementation of effective and sustainable long-term improvements.

Where and how does Six Sigma fit into my business?
A business is essentially a compilation of processes, and Six Sigma is a methodology for improving processes. Consequently, Six Sigma can be deployed in any type of organization and achieve success through process improvement. It can be easily integrated with other initiatives to enhance and build upon improvements that are already in place.

How does Six Sigma apply to different industries?
Six Sigma focuses on process improvement, leading to superior product quality. As a result, Six Sigma is considered a tool for

manufacturing industries only. In fact, nothing could be farther from the truth. Since Six Sigma applies to any type of process, it essentially applies to any organization, regardless of type or size. The effectiveness of Six Sigma is well documented in manufacturing as well as service industries, profits and non-profits, and others. Indeed, the key to success of a Six Sigma project, much as in any other process-focused approach, is the identification of candidate processes for improvement—processes that add value and improve the bottom-line performance. Needless to say, this step is the most crucial and also the most neglected of all the steps involved in the Six Sigma methodology. Some candidate processes for Six Sigma projects are the following:

- Financial transactions
- Customer service
- Order processing
- Product design
- Manufacturing systems

Does Six Sigma apply to nonmanufacturing processes?

Six Sigma has been successfully applied to processes such as billing, insurance claims, customer support, and the like. Needless to say, as in any process, the transactional processes also suffer from process variation, leading to poor quality and longer cycle times. Thus, any such process can be easily improved by applying Six Sigma methodology.

Does Six Sigma apply to administrative operations?

Yes, Six Sigma can be used to optimize administrative operations. It should be noted that in the case of an administrative operation, the main factor affecting system performance is excessive human involvement. Indeed, it is an extremely difficult task to ensure consistently good performance by employees, especially in the absence of an effective measurement system. To complicate the issue further, it is also very difficult to develop robust measurement tools to quantify, control, and manage employees. Since the quality of data in such systems is not itself consistent, this limits the success of the Six Sigma project.

Is Six Sigma applicable to small and midsized companies?
Yes, Six Sigma is applicable to both small and midsized organizations. Although the origins of Six Sigma and its initial successes are linked to large companies, there is no reason why smaller companies shouldn't witness similar proportional successes. Due to its initial successes at Motorola and AlliedSignal, it is widely believed that Six Sigma only benefits big companies, which is totally unfounded. It is important to note that small and midsized companies typically operate at a sigma level of 2 to 3, thereby paying tremendous costs due to poor quality, which they can ill afford. These are the organizations that are affected most severely by poor quality and need help quickly. In addition, from the Six Sigma project standpoint, a smaller company is a better candidate for a successful implementation, since the objectives, scope, and problems are more clearly defined. (Naturally, the number of problems grows exponentially with the size of the organization.) However, due to lack of resources, it is important for a smaller organization to define the project very carefully and keep strict control over the implementation costs.

What types of firms seek Six Sigma performance?
Leading firms in their respective industries have deployed Six Sigma—for example, GE, Motorola, AlliedSignal, Polaroid, and Sony. In fact, Six Sigma is a key differentiator for such companies, enabling them to retain or increase their market dominance today and into the future. As a result of the numerous well-documented successes of market leaders, many companies are starting Six Sigma programs and reporting significant annual savings as a result.

Who is responsible for Six Sigma within an organization or company?
Everyone. Since Six Sigma deals with process control, anyone who manages or works in a process is involved in the imple-

mentation of Six Sigma. Everyone who interacts with a process should learn and practice the tools and techniques to improve that process.

What is a typical Six Sigma organizational architecture?
A successful implementation of the Six Sigma methodology requires the involvement of people from every level of management across the organization. Of course, there are no specific rules for organizing a company to be successful in deploying Six Sigma. As with any other big project, a general rule of thumb is that clearer roles and responsibilities with well-defined objectives and goals at the top will result in a successful implementation. In general, the following roles and responsibilities are typical of any Six Sigma organization:

- *Quality leader (QL)*. Focused on defining the customer requirements to drive better quality and operational effectiveness in the organization.
- *Master Black Belt (MBB)*. Typically assigned to a specific area or function of a business or organization, working with the process owners to ensure that quality objectives and targets are set, plans are defined, progress is tracked, and education is provided.
- *Black Belt (BB)*. Core members of the Six Sigma initiative. Their main purpose is to lead quality projects and work full time until the projects are complete. Black Belts can typically complete 4 to 6 projects per year with savings of approximately $230,000 per project. Black Belts also coach Green Belts on their projects.
- *Green Belt (GB)*. Partially involved in the project while managing their regular work.
- *Process owner (PO)*. Responsible for a specific process.

Why should I give Six Sigma serious consideration?
In the current complex business environment, due to intense competition, no organization can be assured of its lead and market share. The competition is evolving rapidly, and cus-

tomers are bombarded with options. As a result of the recent explosion in the manufacturing and information technologies, the order winners of the past are mere order qualifiers today. To win an order now, companies must excel on multiple dimensions simultaneously, such as the following:

- Quality
- Cost
- Delivery
- Service

As a result, it is of the utmost importance that companies have effective and robust processes to produce competitive products consistently.

Are these the newest and hottest tools?

No, Six Sigma tools are *not* the newest tools. What is new is their application to process control and integration with other programs to create a proactive system for quality control. The use of these tools for delivering value in a comprehensive way is the driving force behind their recent success.

SIX SIGMA TRAINING

What is a Six Sigma Champion?

The Six Sigma Champion is usually a senior executive who owns these projects. The champion is responsible for leading the organization in the direction of quality improvement. The champion is the driving force behind the organization's Six Sigma implementation. Six Sigma Champions are used in many organizations to provide guidance to the program management team. Champions support projects, finalize candidates for process improvements, and manage any administrative and reporting-related activity, to name a few critical activities. But more important, they act as an interface between the project team and senior leadership to quickly resolve any obstacles in the way of a successful implementation.

What comprises a Six Sigma training program?

Typically, the Six Sigma training program includes classroom presentations, discussions, and hands-on training. Typically, the teaching is supplemented with direct application of the tools to an actual project in the students' company.

What are Black Belts and Master Black Belts and what can individuals with these certifications accomplish?

Six Sigma Black Belts are experts in the Six Sigma methodology. They learn and demonstrate proficiency in the Six Sigma methodology and a variety of statistical process control (SPC) techniques that are keys to the methodology. Black Belts drive business results through the implementation of Six Sigma tools on real projects by leading cross-functional process improvement action teams. Typically, Black Belts will work on projects full time for the first 4 to 6 months of their participation in the program. The criteria for attaining Master Black Belt status vary by organization, but in general, these individuals have a few years of experience as a Black Belt, have executed several projects with documented and sustainable results, and have mentored other Black Belt candidates. Master Black Belts act as the organizationwide Six Sigma program manager, overseeing the Black Belts and multiple process improvement projects. They also provide guidance to Black Belts as required. Master Black Belts typically receive additional training in advanced statistical tools, business skills, and leadership skills.

What is required for certification as a Black Belt?

Typically, Black Belts are required to complete two standard projects prior to their certification, of which the first project is completed during the training program. They must also demonstrate understanding and ability to correctly use the Six Sigma tools and methodology.

Do I need to be an expert in statistics to become a Black Belt?

No, expertise in statistics is not a prerequisite to becoming a Black Belt. However, some background in statistics and mathematics will certainly be advantageous.

What is a Six Sigma Green Belt?
Green Belts receive their training from the Black Belts and focus on the DMAIC methodology and intermediate-level quality tools. Black Belts and Green Belts implement the Six Sigma methodology on high-impact improvement opportunities in their organizations to drive prioritized business results. A Six Sigma Green Belt is typically trained less intensively than a Black Belt. In many organizations, Green Belts work on projects part time to support Black Belts.

Does the Six Sigma program build analytical skills?
Six Sigma is based on analysis of data and facts. Actions are taken after thoroughly analyzing the data. After implementing changes, continuous evaluation is performed to ensure that processes are in compliance with the predefined standards. As a result, the program builds analytical and logic skills. In fact, Six Sigma programs are used to train new analysts, who will eventually support future projects in different areas.

Does the Six Sigma approach build future leaders?
Yes, the Six Sigma approach can be used to groom future organizational leaders. Six Sigma programs offer unique opportunities to analyze, understand, and improve large-scale projects comprised of complex processes. Improving such complex systems using a methodical approach with verifiable results in a short period of time requires a well-rounded skill set. After successfully completing such projects, an individual has the following attributes:

• A systemwide view of the business
• Close relations with the top leadership
• A method and tools to improve systems
• Change management experience
• Project management and execution experience

What is the Six Sigma Toolkit?
The standard Six Sigma Toolkit is a collection of techniques and statistical tools that includes the following items:

- Quality function deployment (QFD)
- Process mapping
- Failure mode and effects analysis (FMEA)
- Statistical process control (SPC)
- Regression
- Analysis of variance (ANOVA)
- Design of experiments (DOE)
- Measurement system evaluation (MSE)

SIX SIGMA IMPLEMENTATION

What will it cost to start a Six Sigma program?

It is extremely difficult to estimate the cost of implementing a Six Sigma program without considering the scope and complexity of the implementation. Obviously, the cost of implementing a Six Sigma program is based on many factors. A big component of the total cost is the training cost. However, the training cost is driven by the levels of training and number of candidates receiving the training. In addition, there is the cost of finding replacements to cover the regular tasks of the Black Belt who will be dedicated to the project full time. It is easy to see that if not controlled properly, the Six Sigma program costs can escalate quickly; however, given that the cost of quality for most companies is rather high, the real question is when to start a Six Sigma program.

How do I know which processes in my organization should be targeted for improvement?

A good starting point is to identify the processes that are causing the most problems for the end customers. Based on the criticality of such processes, generate a priority list. It is very important to realize that every process should not be submitted to the improvement project. Due to the limited availability of resources and time, it is important to conduct process selection in a very careful manner. To draw a parallel with the theory of constraints, only the savings at a bottleneck are true savings; the savings at nonbottleneck operations are a mirage.

What resources are required to implement Six Sigma?
The most important resource is the commitment of top management. For effective management of the project, a dedicated leader to drive and manage the implementation is a must. Depending on the scope and objective of the project, key process owners must be involved in the initial phase to define the project scope in detail. Based on the scope, a team of people should be selected for training. This will, in turn, require the availability of experienced trainers—this activity is typically outsourced unless the company is well versed in the Six Sigma tools and technique.

What are the implementation shortcuts for Six Sigma?
The following are key shortcuts:

- Training, training, and more training
- Learning from others' mistakes
- Using people who have successfully implemented it elsewhere
- Using people who understand the potential difficulties
- Planning, planning, and more planning

Will hiring a consultant, a Black Belt, or a Master Black Belt solve our chronic problems and make us a world-class company in 3 or 6 months?
Six Sigma is more than a tool or a method, it is a philosophy. It is critical that the whole organization embraces the fundamentals of Six Sigma and makes it a part of the daily routine. This requires extensive training and nurturing of in-house expertise to promote and sustain a high level of knowledge across the organization. Since Six Sigma projects are done in steps, (i.e., continuous improvement), after the initial support by consultants, all future projects must be driven by internal teams. Consequently, hiring consultants for the first few implementations and to supplement the internal teams in subsequent projects is acceptable. However, using consultants to drive such projects is not a sustainable strategy because of higher costs and the absence of in-house expertise.

Will Six Sigma result in more work for the managers due to extensive data needs?
It is true that the Six Sigma methodology requires more data since it is based on analysis of facts and data. However, streamlined procedures should be developed to collect the necessary data. At the same time, the people affected by the new procedures should be involved from the beginning to analyze and design the process and supporting infrastructure. Once the benefits of the new procedures are realized, it will be easy to maintain the additional burden, if any, due to extensive data needs.

SIX SIGMA RESULTS AND RETURN ON INVESTMENT

How quickly can the investment in Six Sigma be recovered?
The time to recover the investment greatly depends on the scope, complexity, and manner in which the Six Sigma implementation is delivered. A well-scoped and well-managed implementation can be cash-flow positive in as little as 3 to 4 months.

In general, the recovery time is dependent on many factors, namely the following:

- Management sponsorship
- Project selection and alignment of projects to key business objectives
- Proper staffing with the right people
- Realistic schedule
- Proper support

What kinds of business results will the Six Sigma project yield?
A well-executed Six Sigma implementation should deliver significant returns in a short period of time. The main benefit will be improved and consistent processes. As a result of these improvements, the following gains are typical:

- Improved margins and operating costs
- Improved customer satisfaction
- Improved product and service quality/performance
- Reduction in defects
- Reduction of cycle time
- Reduction or elimination of firefighting
- Increased predictability

What drives the magnitude of the business results?
Achievement of business results depends on the following factors:

- Sponsorship
- Project selection
- People
- Training
- Methodology

What type of savings are typical?
The average project savings for a Six Sigma program are approximately $125,000. On average, Black Belts are expected to successfully complete 3 to 4 standard projects per year.

What will my business look like after implementing Six Sigma?
Among other things, the improved business should be:

- More profitable
- More competitive
- More proactive than reactive
- Under control
- More flexible
- More responsive

Will my company achieve the defect rate target of 3.4 ppm (Six Sigma)?

It is important to understand that achieving the defect rate target of 3.4 ppm is not an easy task. The key message of Six Sigma is stepwise defect reduction through continuous process improvements. It is also important for companies to understand that the sustainable optimal defect level varies from one organization to another. Instead of shooting for a standard number, it is better for companies to analyze their data and determine their own targets that are achievable as well as sustainable.

Who has been successful with Six Sigma?

The company with the most publicized Six Sigma success has been General Electric. In its first two years of Six Sigma deployment, GE has saved nearly $900 million, with savings to date of over $1 billion. The person most responsible for promoting Six Sigma at GE was Jack Welch. Other companies that have achieved great success with Six Sigma programs are Navistar (International Truck), AlliedSignal, Motorola, U.S. Postal Service, Honeywell, Nokia, Ford, Gencorp, and Maytag.

How can we calculate the costs and savings for Six Sigma Quality?

A well-planned Six Sigma implementation will lead to balanced costs and savings. A Six Sigma initiative rushed to implementation will lead to poor return on investment due to poor cost tracking and control systems. Interestingly, many companies report savings that resulted from Six Sigma implementation; however, many of them do not report the costs of implementing Six Sigma. For example, GE has invested a little under 0.5 percent of revenue, and realized over 1 percent of revenue during the 5 years of Six Sigma. The following analysis shows estimated cost factors for Six Sigma implementation at different-sized companies.

Number of employees	100	500	1000
Annual sales, $M	10	50	100
Estimated cost of implementing Six Sigma, $M	0.5	1.5	3

Cost of poor quality (COPQ) at 20 percent, $M	2	10	20
Cost of implementing Six Sigma, % COPQ	25	15	15
Minimum number of projects for breakeven	3	10	20

INTEGRATION OF SIX SIGMA WITH OTHER INITIATIVES

How is process sigma improvement related to Total Quality Management and other quality improvement approaches?
The common theme among quality initiatives is that they are customer focused and strive to improve quality by controlling the processes. The essential difference lies in the actual tools and techniques deployed to achieve the results. Compared to other techniques, Six Sigma uses a highly structured methodology and statistical tools to analyze and monitor the process improvement. Another key differentiator is the heavy emphasis placed on team training. The component of training is typically missing from other quality initiatives. As a result of the well-defined structure and the extensive training, Six Sigma projects are generally more successful than similar quality initiatives.

How do Lean manufacturing and Six Sigma relate?
Lean manufacturing and Six Sigma both focus on waste reduction. Lean manufacturing focuses on enabling the fast flow of the product through the manufacturing process with minimal waste. One of the key success factors in achieving this objective is the reduction of variation in various aspects of production. Similarly, Six Sigma methodology also focuses on waste reduction by controlling process variation, quality issues, and training. Clearly, Lean manufacturing and Six Sigma are complementary initiatives and should be deployed together to exploit the resulting synergy.

What differentiates the Six Sigma program from other programs?
The Six Sigma program is a methodical approach that drives improvements that lead to better financial results. If imple-

mented properly, Six Sigma offers guaranteed improvements in a short period of time. The Six Sigma methodology and its tools are tightly interwoven to extend a powerful and complete solution to process problems. For example, the structured approach of Six Sigma requires extensive training to develop in-house expertise, which in turn ensures better results, because the process owners are intimately familiar with the process and can drive performance using the Six Sigma tools. Through extensive training and well-defined projects, the overall character of the organization slowly transforms into that of a learning organization. However, most other quality initiatives are light on the specifics of implementation. As a result, different organizations interpret their opportunities in different manners, and in the absence of a blueprint guiding them through the difficult process of implementation, the end result is an unsuccessful effort.

Why do we need to have a Six Sigma program in addition to ISO 9000 certification?

Although ISO 9000 is aligned with modern quality concepts such as continuous improvement and variation reduction, it has many shortcomings that have limited its usefulness. Since ISO 9000 is an order qualifier in certain business, companies are compelled to obtain ISO 9000 certification. As a result, some companies take a quick-and-dirty approach to achieve ISO 9000 compliance, at least in principle. Obviously, such efforts lack direction and end up as ineffective quality initiatives. However, the Six Sigma initiative is driven from within and guided by customer-focused quality. The key difference lies in the philosophy of the two initiatives. It is thus easy to see that the two initiatives complement each other because of the differences in approach.

How does Six Sigma fit with traditional quality initiatives?

Six Sigma complements all other quality initiatives, such as TQM, ISO 9000, and the like. Most quality initiatives are procedural, organizational, or analytical. In other words, all other initiatives address quality in a very narrow way. Consequently, the solutions also offer limited benefits and are not sustainable.

However, Six Sigma is a comprehensive approach that follows a well-defined proven methodology and deploys effective statistical tools. This results in systems that are continuously improving, yielding superior quality. Six Sigma thus supports and promotes other quality initiatives.

SIX SIGMA BOOKS

Following is a list of most of the Six Sigma–related books on the market. Observations about each book are based on a review of the book by this author or its editorial and customer reviews. Each book has been assigned a category based on my understanding of the book or the emphasis given by the book's author.

Every attempt has been made to prepare a comprehensive list with a fair set of brief observations and assignment of the most suitable topical category. Any error is unintentional. Please notify me so that corrections may be made in future editions.

Following is a list of topical categories of Six Sigma books:

Awareness	Projects
Awareness/methodology	Reference
Case study	Research
Design of experiments	Service—Lean and Six Sigma
DFSS	Six Sigma Scorecard
Handbook	Six Sigma statistics
Lean and Six Sigma	SPC
Methodology/advanced level	Strategic
Pocket guide	

Six Sigma books are sorted by author, title, and year of publication in the following table.

SIX SIGMA BOOKS

Book	Author/Editor	Year of Release	Category	Observations
Six Sigma Deployment	Adams, Cary, Praveen Gupta, and Charlie Wilson	2002	Awareness/ Methodology	This book explains various concepts and the methodology of Six Sigma in practical terms. The book includes case studies and comparison of various quality systems.
The New Six Sigma	Barney, Matt, and Tom McCarty	2002	Awareness	*The New Six Sigma*, a best seller, introduces improvement in the conventional Six Sigma methodology using case studies. The book highlights acceleration, innovation, and a step-by-step approach to the "new" Six Sigma.
The Six Sigma Book for Healthcare	Barry, Robert, Amy Murcko, and Clifford Brubaker	2002	Awareness/ Methodology	This book explains application of Six Sigma in the health-care industry. It includes tips and tools for implementation of Six Sigma.
Quality Beyond Six Sigma	Basu, Ron, and J. Nevan Wright	2003	Awareness/ Methodology	This book combines practical issues of Six Sigma, Lean Enterprise, and Total Quality with softer and strategic issues to develop a derivative called *FIT SIGMA*.
Smart Things to Know About Six Sigma	Berger, Andrew	2003	Awareness/ Methodology	This book shows how to bridge the gap between the theory and the practical applications of Six Sigma. It discusses DMAIC, the project process that teams use for continuous improvement in Six Sigma.
Cause and Effect Lean	Bicheno, John	2000	Awareness/ Methodology	This book covers essentials of Lean manufacturing, including Cause-and-Effect diagrams.

Title	Author	Year	Category	Description
Wisdom on the Green: Smarter Six Sigma Business Solutions	Breyfogle, Forrest, David Enck, Phil Flories, and Thomas Pearson	2001	Awareness/ Methodology	This book explains smarter Six Sigma solutions in a golf game. It explains implementation of Six Sigma, Lean thinking, ISO 9000, Malcolm Baldrige Award, TQM, Balanced Scorecard, Theory of Constraints, and business measurements.
Implementing Six Sigma: Smarter Solutions Using Statistical Methods, Second Edition	Breyfogle, Forrest	2003	Handbook	An excellent reference book on Six Sigma. This book is centered on the author's "Smarter Six Sigma Solutions." It provides an approach, checklists, and metrics for implementation of Six Sigma.
Design for Six Sigma	Brue, Greg	2003	DFSS	This book is a simplified explanation of design for Six Sigma with a checklist at the end of each chapter. It is a part of the series written by the author.
Six Sigma for Managers	Brue, Greg	2002	Awareness/ Methodology	This book introduces Six Sigma to managers. It facilitates better understanding of the concept of Six Sigma, skills improvement, and the commitment required to implement Six Sigma.
The Power of Six Sigma	Chowdhury, Subir	2001	Awareness	This is an excellent book for introducing Six Sigma to managers and executives. This fictionalized tale simplifies a complicated topic and explains the way Six Sigma works in a nonthreatening, easy-to-understand way.
Design for Six Sigma in Technology and Product Development	Creveling, C. M., J. L. Slutsky, and D. Antis	2002	DFSS	This 800-page book links DFSS to various phases of the product development process without losing sight of cycle-time management. It includes step-by-step instructions, flow charts, score cards, and checklists to assist in implementation.

(Continued)

SIX SIGMA BOOKS (*Continued*)

Book	Author/ Editor	Year of Release	Category	Observations
Juran Institute's Six Sigma Breakthrough and Beyond: Quality Performance Breakthrough	De Feo, Joseph, and William Barnard	2003	Strategic	This book goes beyond certification or implementation processes discussed in most Six Sigma texts to prepare an organization's managers. It includes a complete process to improve and maintain Six Sigma breakthrough performance after the initial success.
Six Sigma for Dummies	DeCarlo, Neil, and Craig Cygi	2004	Awareness	Written for Six Sigma beginners, *Six Sigma for Dummies* is a straightforward guide to Six Sigma. This simple and friendly guide makes sense of Six Sigma.
What Makes Six Sigma Work?	Dedeke, A.	2002	Awareness/ Methodology	A research-based book that bring together the key lessons that have been learned about Six Sigma and Lean management, Total Quality Management, and SPC.
Integrating Lean Six Sigma and High-Performance Organizations	Devane, Tom	2003	Lean and Six Sigma	The book describes the integration of Lean and Six Sigma, builds upon their strengths, and addresses the shortcomings of each discipline. The book provides a step-by-step roadmap for integrating Lean and Six Sigma.
Six Sigma Study Guide	Dodson, Bryan, Paul Keller, and Thomas Pyzdek	2001	Reference	A reference book to prepare for the ASQ's Certified Six Sigma Black Belt examination.

Title	Author	Year	Category	Description
Six Sigma for Everyone	Eckes, George	2003	Awareness	An introductory book with definitions and examples for managers and employees. It also explains what Six Sigma is not. The book includes basic tools and techniques for team members, and change management to make Six Sigma work.
The Six Sigma Revolution	Eckes, George	2000	Awareness/ Methodology	This book addresses various aspects of Six Sigma implementation in a corporation. It includes well-defined methods for launching the Six Sigma initiative.
Lean Six Sigma for Service	George, Michael	2003	Service—Lean and Six Sigma	This best-selling author applies his innovative approach of integrating Lean and Six Sigma to the service environment. The book provides examples of the critical determinants of quality and speed for the flow of information and the interaction between people.
Lean Six Sigma	George, Michael	2002	Lean and Six Sigma	This best-selling book provides an innovative approach to integrating Lean and Six Sigma concepts. The author presents a step-by-step roadmap for profiting from the best elements of Lean and Six Sigma.
What Is Lean Six Sigma?	George, Michael, David Rowlands, and Bill Kastle	2003	Awareness/ Methodology	The book summarizes the concepts of Lean Six Sigma. It demonstrates why companies are implementing Lean Six Sigma, and benefiting from doing so.
Six Sigma Quality for Business and Manufacture	Gordon, Joseph	2002	Methodology/ Advanced level	This book includes detailed checklists, questionnaires, and forms to assist people in preventing problems from occurring in first place, and to solve new- and long-term problems in services and manufacturing.

(Continued)

Book	Author/ Editor	Year of Release	Category	Observations
Six Sigma Business Scorecard	Gupta, Praveen	2003	Six Sigma Scorecard	This book establishes a new performance measurement system that relates to corporate growth and profitability. The Six Sigma Business Scorecard is developed by combining Six Sigma and Balanced Scorecard. The book also defines a model for calculating corporate Sigma level. This is an excellent book for establishing the right measurements to sustain the Six Sigma initiative.
Six Sigma: The Breakthrough Management Strategy Revolutionizing the World's Top Corporations	Harry, Mikel, and Richard Schroeder	1999	Strategic	This is an excellent Six Sigma introduction for executives. It provides an overview of Six Sigma with enthusiasm and conviction, owing to the personal accomplishments of the authors. Harry and Schroeder explain Six Sigma and show how it has worked at companies such as General Electric, Polaroid, and Allied Signal. The authors show how most companies work at a "sigma" level of between 3.5 and 4 without a Six Sigma initiative.
Demystifying Six Sigma	Larson, Alan	2003	Awareness/ Methodology	This book helps readers to gain a clear understanding of customer needs and what is required for success through application of the legendary Quality Assurance program across all departments and processes. This creates a permanent, companywide Six Sigma culture.

Title	Author	Year	Category	Description
Deploying Six Sigma to Bolster Business Processes and the Bottom Line	Leavitt, Paige	2001	Research	his book is based on the research and resources of APQC (American Productivity and Quality Center). It describes Six Sigma as a methodology to help companies meet the needs of their business.
Achieve Lasting Process Improvement: Reach Six Sigma Goals Without the Pain	Lientz, Bennet, and Kathryn Rea	2002	Awareness/ Methodology	This book focuses on process improvement through Six Sigma, offering practical tips and guidelines for specific issues.
Defining and Analyzing a Business Process	Lowenthal, Jeffrey	2002	Pocket Guide	This pocket guide focuses on methodology to analyze existing processes for improvement.
Six Sigma Tool Navigator	Michalski, Walter, and Dana King	2003	Handbook	The book contains tools, tables, and illustrations used in implementing Six Sigma. It is a guide to assist teams in their problem-solving efforts.
Leaning into Six Sigma: The Path to Integration of Lean Enterprise and Six Sigma	Mills, Chuck, Barbara Wheat, and Mike Carnell	2001	Lean and Six Sigma	This is an introductory book to read before starting Lean/Six Sigma deployment. The authors explain in a non-technical manner the Six Sigma problem-solving methodology and link it to Lean principles.
Six Sigma for the Shop Floor	Munro, Roderick	2001	Pocket Guide	This book is written to introduce Six Sigma to line workers. It explains the principles to operators and helps them with Sigma projects.
Guiding Successful Six Sigma Projects	Oriel, Inc.	2000	Projects	This spiral-bound book summarizes various activities and roles with projects for Managers, Project Sponsors, Master Black Belts, and Black Belts. The book highlights various check-lists to be used during a meeting or to monitor a project.

(Continued)

SIX SIGMA BOOKS (*Continued*)

BOOK	AUTHOR/ EDITOR	YEAR OF RELEASE	CATEGORY	OBSERVATIONS
What Is Design for Six Sigma?	Pande, Peter, Bob Neuman, and Roland Cavanagh	2004	Awareness/ Methodology	This book reveals how to use DFSS to design new products, services, and processes so that quality problems can be prevented.
The Six Sigma Way	Pande, Peter, Robert Neuman, and Roland Cavanagh	2000	Awareness/ Methodology	The book, a best seller, provides a comprehensive guide to applications of Six Sigma. It focuses on a five-step "roadmap" tied to a company's areas or processes.
The Six Sigma Way Team Fieldbook	Pande, Peter, Robert Neuman, and Roland Cavanagh	2002	Handbook	This book mainly focuses on the project improvement teams for hands-on work. It offers improvement guidelines using advanced Six Sigma tools.
The Six Sigma Handbook, Second Edition	Pyzdek, Thomas	2003	Handbook	This book includes a comprehensive set of statistical and quality improvement tools used in the Six Sigma methodology. It has added modules of DFSS and FMEA and shows how to avoid pitfalls. The book covers advanced topics such as DOE and RSM.
Black Belt Memory Jogger: Pocket Guide for 6 Sigma Success	QPC, Goal	2002	Pocket Guide	This is a spiral-bound detailed reference book for Six Sigma practitioners. It includes a section on how to use Six Sigma tools, which can be used by practitioners while solving problems.

Title	Author	Year	Category	Description
Rath & Strong's Six Sigma Champion Pocket Guide	Rath & Strong	2003	Pocket Guide	This is a guide advising Six Sigma Champions how to apply DMAIC methodology appropriately.
Rath & Strong's Six Sigma Pocket Guide	Rath & Strong	2000	Pocket Guide	This is a good pocket reference guide that describes DMAIC tools and how they are used.
Rath & Strong's Six Sigma Team Pocket Guide	Rath & Strong	2003	Pocket Guide	This book focuses on team processes for Six Sigma projects.
Blending Lean and Six Sigma Workbooks	Rubrich, Larry, Madelyn Watson, and Alan Larson	2001	Lean and Six Sigma	This spiral-bound book offers comprehensive guidelines for integrating and implementing tools such as 5S, Teams, TPM, Cells Operations, Setup Reduction, and Kanbans.
Strategic Six Sigma: Best Practices from the Executive Suite	Smith, Dick, and Jerry Blakeslee	2002	Awareness/ Methodology	This book looks at the strategic intent of Six Sigma and offers a roadmap for its implementation. The book incorporates best practices for better understanding of Six Sigma at the executive level.
Leading Six Sigma	Snee, Ronald D., and Roger W. Hoerl	2002	Strategic	The book provides a systematic approach for launching a Six Sigma initiative in an organization. Written specifically for corporate leaders with a clear view from the top, and it provides start-to-finish deployment of Six Sigma.
Six Sigma and Beyond: Design of Experiments	Stamatis, D. H.	2002	Design of Experiments	This author focuses on the Design of Experiments (DOE) aspects of Six Sigma. The book covers topics from the concept of variation all the way to robustness.

(Continued)

SIX SIGMA BOOKS (Continued)

Book	Author/Editor	Year of Release	Category	Observations
Six Sigma and Beyond: Foundations of Excellent Performance	Stamatis, D. H.	2001	Awareness/ Methodology	This volume focuses on the basics of Six Sigma, its impact on cost reduction, productivity improvement, customer retention, and enhanced bottom line. It answers frequently asked questions about Six Sigma.
Six Sigma and Beyond: Problem Solving and Basic Mathematics	Stamatis, D. H.	2001	Six Sigma Statistics	This volume provides a generic approach to problem solving. It addresses key concepts of teamwork and basic concepts of problem solving. The book highlights the basic mathematical concepts of Six Sigma with examples.
Six Sigma and Beyond: Statistical Process Control	Stamatis, D. H.	2002	SPC	This book primarily focuses on SPC methodology to reduce variation.
Six Sigma and Beyond: The Implementation Process	Stamatis, D. H.	2002	Reference	This book highlights the curriculum required for a Six Sigma program. The author provides implementation guidelines for various tools and techniques covered in the preceding volumes.
Six Sigma for Financial Professionals	Stamatis, D. H.	2003	Awareness/ Methodology	This book provides guidelines for implementing Six Sigma methodology in the financial industry using many examples.

Title	Author	Year	Type	Description
Six Sigma Fundamentals	Stamatis, D. H.	2003	Awareness/ Methodology	This guide gives an overview to the entire process of implementing Six Sigma. The book includes a CD-ROM with many forms and provides the reader with a solid understanding of what defines a Six Sigma initiative and what is expected from the organization, management, and customer.
Understanding the Essentials of the Six Sigma Quality Initiative	Star, Harold, and Stephen Snyder	2000	Awareness	This 100-page book explains concepts of Six Sigma. It's useful for communicating the Six Sigma initiative and its importance to employees who are not directly involved, but who need to be on board to imbue it into the corporate culture.
Six Sigma and Related Studies in the Quality Disciplines	Stephens, Kenneth	2003	Awareness	This book offers a collection of articles about Six Sigma from an international perspective. The articles cover topics that include reduction of variation, strategies for implementing Six Sigma, Six Sigma in service industries, and limitations of Six Sigma.
Six Sigma: SPC and TQM in Manufacturing and Services	Tennant, Geoff	2001	Awareness	The book establishes the relationships among Six Sigma, quality, customer satisfaction, business processes, organizational structure, statistics, analysis, and process improvement methodologies.
Weight Loss Success Using Six Sigma	Thompson, Sterling	2003	Case study	The book demonstrates how Six Sigma can be used to lose weight and stay lean.

(*Continued*)

SIX SIGMA BOOKS (*Continued*)

Book	Author/ Editor	Year of Release	Category	Observations
The Six Sigma Path to Leadership: Observations from the Trenches	Treichler, David, and Ronald Carmichael	2004	Awareness	This book is written to inspire and motivate by using Six Sigma through stories, and "how" and "how not to" examples. It is written for anyone involved with the Six Sigma initiative.
The Six Sigma Journey from Art to Science	Walters, Larry	2002	Awareness	This is an unusual book that introduces the Six Sigma methodology through real-life examples. Through story telling, the author conveys the technical as well as human aspects of implementing Six Sigma successfully.
Leaning Into Six Sigma	Wheat, Barbara, Chuck Mills, and Mike Carnell	2003	Lean and Six Sigma	This book is a business novel that tells the story of implementing a Lean Six Sigma initiative and improving corporate performance.
Design for Six Sigma: A Roadmap for Product Development	Yang, Kai, and Basem S. El-Haik	2003	DFSS	This volume provides a walk-through that helps the reader choose the best design tools at every stage of the development process. It gives a roadmap to produce excellence through designing for Six Sigma.

BIBLIOGRAPHY

Adams, Cary W., Praveen Gupta, and Charles E. Wilson, Jr. *Six Sigma Deployment.* Burlington: Butterworth-Heinemann, 2002.

Anderson, Dave. *Up Your Business.* Hoboken, NJ: Wiley, 2003.

Altshuller, G. *And Suddenly the Inventor Appeared: TRIZ, the Theory of Inventive Problem Solving.* Worcester, MA: Technical Innovation Center, 1996.

Altshuller, G. *40 Principles: TRIZ Keys to Technical Innovation.* Worcester, MA. Technical Innovation Center, 1997.

Barney, Matt, and Tom McCarty. *The New Six Sigma,* Upper Saddle River, NJ: Prentice Hall, 2003.

Bhote, Keki R. *The Power of Ultimate Six Sigma.* New York: Amacom, 2003.

Box, G. E. P., W. G. Hunter, and J. S. Hunter. *Statistics for Experimenters.* New York: Wiley, 1978.

Breyfogle, Forrest W. III. *Implementing Six Sigma: Smarter Solutions Using Statistical Methods.* New York: Wiley, 1999.

Brue, Greg. *Design for Six Sigma,* New York: McGraw-Hill, 2003.

Brue, Greg. *Six Sigma for Managers.* New York: McGraw-Hill, 2003.

Chowdhury, Subir. *The Power of Six Sigma.* Chicago, Dearborn, 2001.

Christensen, Clayton M. *The Innovator's Dilemma.* New York: HarperBusiness, 2003.

Christensen, Clayton M., and Michael E. Raynor. *The Innovator's Solution.* New York: Harvard Business School Press, 2003.

Collins, Jim. *Good to Great.* New York: HarperCollins, 2002.

Collins, Jim, and Jerry Porras. *Built to Last.* New York: HarperCollins, 1994.

Dauphinais, G. William, and Colin Price (eds.). *Straight from the CEO: The World's Top Business Leaders Reveal Ideas That Every Manager Can Use.* New York: Simon & Schuster, 1998.

DeBono, Ed. *Surpetition.* New York: Harper Business, 1992.

DeFeo, Joseph A., and William W. Barnard. *Six Sigma Breakthrough and Beyond.* New York: McGraw-Hill, 2003.

Deming, W. Edwards, *Out of the Crisis.* Cambridge, MA: MIT Press, 1988.

Duncan, Ronald, and Miranda Weston-Smith (eds.). *The Encyclopedia of Ignorance.* New York: Pocket Books, 1977.

Eckes, George. *Six Sigma for Everyone.* Hoboken, NJ: Wiley, 2003.

Edvinsson, Leif, and Michael S. Malone. *Intellectual Capital: Realizing Your Company's True Value By Finding Its Hidden Brainpower.* New York: HarperBusiness, 1997.

Finkelstein, Sydney. *Why Smart Executives Fail*. New York: Penguin Group, 2003.

Garner, Joe, with narration by Bill Kurtis. *We Interrupt This Broadcast*. Naperville, IL: Sourcebooks, 1998.

Garvin, David. *Building a Learning Organization*. Cambridge, MA: Harvard, 1993.

George, Michael L. *Lean Six Sigma: Combining Six Sigma Quality with Lean Speed*. New York: McGraw-Hill, 2002.

Gladwell, Malcolm. *The Tipping Point: How Little Things Can Make a Big Difference*. Boston: Little, Brown, 2000.

Graham, Jacqueline D., and Michael J. Cleary (eds.). *Problem Solving and Planning Tools: Practical Tools for Continuous Improvement*, vol. 1. Miamisburg, OH: PQ Systems, 1992–2000.

Graham, Jacqueline D., and Michael J. Cleary (eds.). *Statistical Tools: Practical Tools for Continuous Improvement*, vol. 2. Miamisburg, OH: PQ Systems, 1992–2000.

Groebner, D. F., P. W. Shannon, P. C. Fry, and K. D. Smith. *Business Statistics: A Decision Making Approach*, 5th ed., Upper Saddle River, NJ: Prentice Hall, 2001.

Gupta, Praveen. *Ask the Expert*. iSixSigma.com, February 2004.

Gupta, Praveen. *ISO 9000:200 An Implementation Guide*. IL: Quality Technology Company, 2001.

Gupta, Praveen. *Six Sigma Business Scorecard*. New York: McGraw-Hill, 2003.

Hammer, Michael, and Steven A. Stanton. *The Reengineering Revolution*. New York: HarperCollins, 1995.

Harry, Mikel. *Six Sigma Knowledge Design*. Palladyne, 2001.

Harry, Mikel. *The Vision of Six Sigma*. Phoenix, AZ: Star, 1997.

Harry, Mikel, and Richard Schroeder. *Six Sigma: The Breakthrough Management Strategy Revolutionizing the World's Top Corporations*. New York: Doubleday, 2000.

Harvard Business Review on Knowledge Management. New York: Harvard Business School Press, 1987, 1991, 1993, 1996–1998.

Juran, J. M. *Juran's Quality Handbook*. New York: McGraw-Hill, 1998.

Juran, J. M., and F. M. Gryna, Jr. *Quality Planning and Analysis*, 2nd ed. New York: McGraw-Hill, 1980.

The Juran Institute. *The Six Sigma Basic Training Kit*. New York: McGraw-Hill, 2001.

Kaplan, Robert S., and David P. Norton. *The Balanced Scorecard: Translating Strategy Into Action*. New York: Harvard Business School Press, 1996.

Klibanoff, P., B. Moselle, and A. Sandaroni. *Statistical Methods and Managerial Decisions Case Pack*. Kellogg School of Management, Northwestern University, 2000.

Labovitz, George, and Victor Rosansky. *The Power of Alignment: How Great Companies Stay Centered and Accomplish Extraordinary Things.* New York: Wiley, 1997.

LaBrake, Mary L. *Tests for Differences.* Reading, MA: Addison-Wesley, 1992.

Lencioni, Patrick. *The Five Disfunctions of a Team.* San Francisco: Jossey-Bass, 2002.

Lubben, Richard. *Just-In-Time Manufacturing.* New York: McGraw-Hill, 1988.

McPherson, James M. *Battle Cry of Freedom: The Civil War Era.* New York: Ballantine, 1998.

Montgomery, Douglas C. *Design and Analysis of Experiments.* New York: Wiley, 1976.

Naumann, Earl, and Steven H. Hoisington. *Customer Centered Six Sigma: Linking Customers, Process Improvement and Financial Results.* Milwaukee, WI: ASQ Quality Press, 2001.

Nonaka, Ikujiro. *The Knowledge Creating Company.* New York: Oxford, 1995.

Pande, Peter S., Robert P. Neuman, and Roland R. Cavanagh. *The Six Sigma Way: An Implementation Guide for Process Improvement Teams.* New York: McGraw-Hill, 2002.

Phadke, Madhav S., *Quality Engineering Using Robust Design.* Englewood Cliffs, NJ: Prentice Hall, 1989.

Pyzdek, Thomas. *The Six Sigma Handbook: A Complete Guide for Greenbelts, Blackbelts, and Managers at All Levels.* New York: McGraw-Hill, 2001.

Rother, Mike, and Rick Harris. *Creating Continuous Flow.* Lean Enterprise Institute, 2002.

Rother, Mike, and John Shook. *Learning to See.* Lean Enterprise Institute, 1998.

Shewhart, Walter. "Economic Control of Quality of Manufactured Product." 1931.

Smith, Dick, and Jerry Blakeslee. *Strategic Six Sigma.* Hoboken, NJ: Wiley, 2002.

Stapleton, James J. *Executive Guide to Knowledge Management.* Hoboken, NJ: Wiley, 2003.

Stewart, Thomas A. *Intellectual Capital, The New Wealth of Organizations.* New York: Doubleday/Currency, 1997.

Sullivan, Gordon, and Michael Harper. *Hope Is Not a Method: What Business Leaders Can Learn from America's Army.* New York: Time Business/Random House, 1996.

Ulrich, Dave, and Norm Smallwood. *Why the Bottom Line Isn't! How to Build Value through People and Organizations.* Hoboken, NJ: Wiley, 2003.

Welch, Jack. *Jack: Straight from the Gut.* New York: Warner Business Books, 2000.

Wheat, Barbara, Chuck Mills, and Mike Carnell. *Leaning into Six Sigma.* New York: McGraw-Hill, 2003.

Womak, James. *Lean Thinking: Banish Waste and Create Wealth in Your Organization.* New York: Free Press, 1996.

Womack, James. *The Machine That Changed the World.* Perennial, 1991.

Zook, Chris, with James Allen. *Profit from the Core: Growth Strategy in an Era of Turbulence.* New York: Harvard Business School Press, 2001.

WEBSITES

www.apssa.uiuc.edu

www.dnh.mv.net

www.emsstrategies.com

www.improvComedy.org

www.innovationtools.com

www.innovation-triz.com

www.isixsigma.com

www.mamtc.com

www.manex.org

www.mindtools.com

www.mit.edu

www.mycoted.com

www.powerdecisions.com

www.processmaps.com

www.scu.edu.au

www.sytsma.com

www.teamtechnology.co.uk

www/theaccessgroupllc.com

www.tocforme.com

www.tpmonline.com/

www.triz.org

www.triz-journal.com

INDEX

NOTE: **Boldface** numbers indicate illustrations.

ABOUT THE AUTHOR

Praveen Gupta, a Master Six Sigma Black Belt, has been involved with Six Sigma since 1996. His first project, completed at Motorola under the guidance of Bill Smith, the inventor of Six Sigma, was published in *Quality Engineering* magazine in 1989. For over 15 years, Praveen has published articles, monthly columns, and books on Six Sigma. His previous books include *Six Sigma Deployment* and *Six Sigma Business Scorecard*. Praveen has conducted Six Sigma training worldwide since 1987 and has assisted others in their projects to solve their problems creatively.

Praveen has consulted with about 100 small- to medium-sized companies in corporate performance improvement through Six Sigma and business process improvement. Praveen is a co-chairman of IPC's SPC committee, which has recently developed an industry standard, IPC-9194, for process controls. Praveen is a Fellow of ASQ. He lives in Lisle, Illinois.

ABOUT THE CONTRIBUTORS

Patricia (Patty) Barten is a former Global Division General Manager and Vice President at Motorola. She has been recognized as a catalyst for comprehensive advancement organizational performance, which encompasses profits, customers, people, processes, systems, and culture objectives. She is differentiated by her inspirational, visible, and invested leadership style coupled with adaptable business acumen. Patty directed one of the first Six Sigma projects in the electronics manufacturing area at Motorola in 1988 and received the CEO Award for her area.

Kam Gupta is President of Continuous Improvement Technology, Inc., Buffalo Grove, Illinois, where he supports his clients' process and quality improvement initiatives through knowledge-based change. Kam holds an MSEM from the Milwaukee School of Engineering. Kam Gupta has been a member of the Board of Examiners for the Malcolm Baldridge National Quality Award and is an active member of the American Society for Quality (ASQ).

Marjorie Hook begun her Six Sigma training and consulting career at Motorola in 1990. During the years she worked with Motorola she had the opportunity to explore Six Sigma applications in each of Motorola's manufacturing sectors, as well as with key customers and suppliers. Since that time she has also participated in implementations across high technology service, banking, and healthcare. Her latest ventures are in the rapidly evolving Lean Six Sigma space.

Priya Ponmudi, an industrial engineer, works at a door and window manufacturing company. She holds an M.S. degree in manufacturing and information systems from Florida International University. Her job experience includes extensive reprocessing of existing manufacturing processes, feasibility studies, and product improvements. She assisted in a Six Sigma project at Amex that saved about $3,000,000.

Shan Shanmugham is an ASQ-certified Six Sigma Black Belt with over 20 years of experience in continuous improvement efforts. Shan has developed processes for new product management and process improvement efforts in the manufacturing and services sector. He holds a MSIE in operations research from the Wayne State University, and an MBA from Michigan State University, and holds several product and process patents. His current interests include knowledge management, closing/pricing of services, and body wearable computing in teleservicing.

Mahender Singh has worked for over 8 years in the area of process management and planning. Mahender is well-versed in the art of modeling business problems and exploiting the power of statistical data analysis and mathematical optimization. He has earned advanced degrees in the fields of business administration (Ph.D.), management science (M.S.), statistics (M.S.), and accounting. His current research interests are in the area of logistics and supply chain management, and queuing optimization techniques.

Arvin Srivastava has about 15 years of experience in business process improvement in several industries. He holds a BSME degree from IIT Roorkee and an MBA degree from Benedictine University, Illinois. He is a Six Sigma Black Belt and is employed at Quality Technology Company.

Rajesh Tyagi is an adjunct faculty member in the Department of Managerial Economics and Decision Sciences of the Kellogg School of Management, Northwestern University. His primary research interests are in the areas of optimal service chain design, the role of product management in hi-tech industry, and supply-chain risk management. Dr. Tyagi obtained his Ph.D. in process engineering at the University of Ottawa and his MBA at the Kellogg School of Management, Northwestern University.

Rajiv Varshney is a senior performance analyst with Household Card Services. He has over 8 years of experience in Six Sigma, supply chain, and operations management with Ernst & Young Consulting and General Motors, among others. Rajiv has an MBA from Michigan State University and is an ASQ Certified Six Sigma Black Belt.